Mountains Without Handrails

Mountains Without Handrails

Reflections on the National Parks

Joseph L. Sax

Ann Arbor The University of Michigan Press

For Mary Sax

Published in the United States of America by
The University of Michigan Press
Manufactured in the United States of America
♾ Printed on acid-free paper

2001 18 17 16

Grateful acknowledgment is made to Jerry Schurr, the artist, John Szoke Graphics Inc., the publisher, and Original Print Collectors Group, Ltd. for permission to use "Yosemite" as the cover illustration. Copyright © by Jerry Schurr.

Library of Congress Cataloging in Publication Data

Sax, Joseph L
Mountains without handrails, reflections on the national parks.

Bibliography: p.
Includes index.
1. National parks and reserves—United States.
I. Title.
E160.S29 1980 917.3 80-36859
ISBN 0-472-06324-3 (pbk.)

Acknowledgments

I began this book during my tenure as a fellow at the Center for Advanced Study in the Behavioral Sciences at Stanford, an ideal place to work. My time there was in part made possible by a grant from the Ford Foundation and the understanding helpfulness of Marshall Robinson and William Pendleton at the foundation. I have also had the aid of the William W. Cook Fund at the University of Michigan Law School.

I am grateful to friends and colleagues, who read my manuscript and provided critical suggestions: Phil Soper, Fred Small, Elli Sax, Terry Sandalow, Don Regan, Spense Havlick, and Lee Bollinger.

Contents

Introduction

I have long been fascinated by the political influence of that small minority who—for lack of a more exact term—are generally known as preservationists.[1] In good times and bad, for over a century, they have regularly persuaded the Congress to establish and maintain national parks, insulating millions of acres of public land not only from most commercial and industrial use, but even from much of the development that popular tourism demands. During the heyday of utilitarian forestry eighty years ago, they were called "nature fakirs," a cruel joke that expresses almost perfectly the ambivalence of the majority toward the politics of preservation.

The public greatly admires splendid scenery and untrammeled nature, as frequent television specials, magazine articles, and large sales of coffee table picture books attest; and it nods in agreement at a steady flow of press reports, all more or less entitled "Are We Loving Our National Parks to Death?" At the same time there is widespread frustration and resentment when—at the behest of the "nature fakirs"—government refuses to build roads into the wilderness, to accommodate more recreational vehicles in the parks, or to approve an elegant ski resort in an alpine valley.

The preservationist is in rather the same position as the scientist who comes to the government seeking research funds. He speaks for something most people admire without understanding, receives unstinting support for a while, only suddenly to be turned upon by a wave of popular reaction against alleged elitism and arrogance.

Whatever the problems of scientific researchers, it is at least recognized that they know something beyond the ken of most of us, and that somehow what they are doing is important. The preservationist is not quite so fortunate. It isn't at all obvious that he knows anything special. Attitudes toward nature and recreational preferences seem purely matters of private taste. The auto tourist sees himself as every bit as virtuous as the backpacker. The preservationist often appears as nothing more than the voice of effete affluence, trying to save a disproportionate share of the public domain for his own minoritarian pleasures.

Since the preservationist does not seem to speak for the majority and its preferences, at least in much of what he advocates, on what basis does he come to government, seeking official status for his views? Is he, like the scientist or even the museum director or university professor, the bearer of a great cultural or intellectual tradition? Is he a spokesman for minority rights, or diversity, seeking only a small share of our total natural resources? Or is he the prophet of a secular religion—the cult of nature—that he seeks to have Congress establish?

It may seem odd to be raising such questions more than a century after the first national parks were established. It is my thesis that preservationist ideology—though it has never gone unquestioned—long found itself compatible with a number of other popular desires that our parklands served, and therefore never received the scrutiny or the skepticism to which it is now being subjected. The enormous growth of recreation in recent years and the vastly increased range and mobility of large numbers of tourists has brought long-somnolent questions to the surface. Should the national parks[2] basically be treated as recreational commodities, responding to the demands for development and urban comforts that visitors conventionally bring to them; or should they be reserved as temples of nature worship, admitting only the faithful?

Strictly speaking, these are questions that the Congress answers, for it makes the laws that govern the public lands. They are issues to which the National Park Service must respond on a daily basis, for it is the bureaucracy that manages these lands. But neither of these two public institutions operates in a vacuum. Both respond to leadership elites that claim to speak legitimately for important public values; and both are sensitive to the limits of public tolerance for self-appointed leaders of opinion. For this reason I propose to ask how the preservationist justifies his asserted leadership, and why—if at all—the public should be inclined to follow.

1

Quiet Genesis

In the last decades of the nineteenth century the federal government began to set aside—out of the vast public domain it was giving away to settlers, railroad companies, and the states—large areas of remote and scenic land to be held permanently in public ownership and known as national parks.[1] What exactly was meant to be accomplished by these unprecedented reservations is a mystery that will never be fully solved. There was at the time no tradition of rural nature parks anywhere in the world.[2] Neither was there a popular movement calling for the establishment of such places,[3] and the first park—the Yosemite Valley and the nearby Mariposa Grove of big trees in California—was created during the Civil War without fanfare, with hardly any congressional debate, and with a minimum of public notice.[4]

The quiet genesis of the national park system is hardly surprising, for the western mountain lands were then virtually unknown. To reach Yosemite Valley in the 1860s, it was necessary to take a boat from San Francisco to Stockton, followed by a sixteen-hour stagecoach ride to Coulterville, and finally a fifty-seven-mile, thirty-seven-hour trek by horse and pack mule into the valley.[5] Yellowstone, established in 1872, was even less accessible. Except to a handful of pioneers, it was unexplored territory, and reports of its spectacular thermal features were widely disbelieved as the inventions of mountain tale spinners.[6] Nor were those who urged the Congress to reserve these places celebrated figures in American life. The Yosemite bill was introduced on the basis of a letter to a California senator from a man named Israel Ward Raymond, described only as a gentleman "of fortune, of taste and of refinement," and of whom all that is known is that he was the California representative of the Central American Steamship Transit Company.[7] The popular account of Yellowstone's founding holds that the idea for a park was conceived by one of the early exploratory parties in the area at an after-dinner campfire in 1870 which decided that so wonderful a region ought never to be allowed to fall into private ownership. Scholarly research has turned up a more plausible, if less romantic, story.[8] One A. B. Nettleton, an agent for the Northern Pacific Railroad Company, passed on to Washington a suggestion which struck him "as being an excellent one. . . . Let Congress pass a bill reserving the Great Geyser Basin as a public park forever. . . ."[8] Subsequently the Northern Pacific became the principal means of access to Yellowstone and its first concessioner providing services for tourists.

The statutes setting aside the first national parks were as cryptic as their histories. Yosemite was turned over to the state of California, to be withdrawn from settlement and held "for public use, resort and recreation."[9] Years later, it was returned to the United States and added to the much larger surrounding lands that comprise most of the present national park. Eight years after the Yosemite grant, Congress similarly withdrew Yellowstone from settlement and dedicated it "as a public park or pleasuring ground for the benefit and enjoyment of the people."[10] In the decades that followed, using a similar verbal formula, Congress set aside Sequoia and Kings Canyon (originally known as General Grant Park) in the high mountains of California, Crater Lake in Oregon, Washington's glacier-capped Mount Rainier, the Indian ruins at Mesa Verde in Colorado, and a number of other remarkable places.[11] It even made Michigan's Mackinac Island a national park in 1875, only to repent and relinquish it three years later. In the first years of the twentieth century it added obscurity to magnificence by adding Wind Cave and Sully's Hill national parks in the Dakotas and Platt National Park in Oklahoma.

If the government had a plan for the parks it was establishing, it was certainly casual about it. No bureau existed to manage these places until 1916, forty-four years after the Yellowstone reservation.[12] Yellowstone, in fact, was run by the United States Cavalry, and the others were pretty much left to themselves and to a few hardy innkeepers and adventurous tourists.[13] The modern desire to view the parks as the product of a prophetic public ecological conscience has little history to support it. The early parks were reserved for their scenery and their curiosities, and they reflect a fascination with monumentalism as well as biological ignorance or indifference.[14]

The ability of a national park system to come into being and to persist most likely grew out of the happy convergence of a number of very diverse, but compatible, forces. Proposals to preserve scenic places followed a period of romantic idealism that had swept the country—the religious naturalism of Thoreau and Emerson, romanticism in the arts, and early nostalgia for what was obviously the end of the untamed wilderness, already in submission to the ax, the railroads, and the last campaigns against the Indians.

The rapidity and relentlessness of settlement also gave weight to efforts to reserve these remarkable sites. When the first Yosemite bill was put before Congress in 1864, the principal claim made was that reservation was necessary to prevent occupation of the valley by homesteaders and to preserve its trees from destruction.[15] Not many years later, John Muir worked for an enlargement of the park to protect the high valleys from the destructive grazing of sheep which he called "hoofed locusts."[16]

Spectacular scenery brought out curiosity seekers eager to turn wonders into profits. As early as 1853 some promoters denuded a number of large sequoia trees of portions of their bark, which were shipped to London to be exhibited for a fee. Ironically, the size of the trees from which the bark came was, to Europeans, so large as to be beyond belief, and the exhibition, thought to be a fraud, was a financial failure.[17] Souvenir hunters were also on the scene, and even early reports from Yellowstone remarked that "visitors prowled around with shovel and ax, chopping and hacking and prying up great pieces of the most ornamental work they could find; women and men alike joining in the barbarous pastime."[18]

Ruthless exploitation of natural marvels stimulated an uneasiness that was felt more generally about the burgeoning spirit of enterprise in the country. Houses were going up, and trees coming down, with such unbridled energy that it was easy to wonder whether Americans valued anything but the prospect of increased wealth. Thoreau's metaphor of lumbermen murdering trees was invoked repeatedly.[19] Andrew Hill, who led the effort to establish the Big Basin Redwood Park in California, is said to have formed his resolve when the private owner boasted that he planned to fell ancient redwoods on his land for railroad ties and firewood. An article in the *Overland Monthly* magazine, urging establishment of a Big Basin park, described the principal enemy of the redwoods not as fire, but as "the greed, the rapacity, the vandalism that would hack and cut and mutilate the grandest, the most magnificent forest that can be found on the face of the earth."[20]

The idea of publicly held parks was not only a predictable response to despoliation and avarice, it also harmonized with a principle that was then at the very crest of its influence in American land policy. The Yellowstone-Yosemite era was also the time of Homestead and Desert Land acts, when every American family was to have its share of the public domain free of monopolization by the rich.[21] The application of that principle to the great scenic wonders could not be realized by granting a sequoia grove or Grand Canyon to each citizen. But it was possible to preserve spectacular sites for the average citizen by holding them as public places to be used and enjoyed by everyone. The fear of private appropriation was far from hypothetical. In 1872, the same year that Yellowstone was established, an English nobleman named Windham Thomas Wyndam-Quin, the fourth earl of Dunraven, came to Colorado on a hunting trip, visiting the area where Rocky Mountain National Park is now located. He casually announced that he wanted to acquire the whole region as a private hunting preserve, and by enlisting a cadre of drifters to file homestead claims for him he was able to gain control of more than fifteen thousand acres. Fortunately, as it happened, the Wild-West style was still in force, and local people, under the leadership of a colorful character known as Rocky Mountain Jim, made things more than a little uncomfortable for Dunraven, who thought he could transpose the style of the European aristocrat to the Colorado mountains. By 1907, Lord Dunraven wrote in his memoirs, he had "sold what [I] could get and cleared out, and I have never been there since."[22]

The park concept also fitted neatly with the nationalistic needs of the time. It appealed to a tenacious American desire to measure up to European civilization. What little discussion one finds in early congressional debates is full of suggestions that our scenery compares favorably to the Swiss Alps and that we can provide even more dazzling attractions for world travelers.[23] In the awesome scenery of the mountain West, America had at last a way to compete on an equal plane with the Old World. This prospect was not lost on the railroads, then the most important element in the growing tourist industry, and their support for national parks was never far beneath the surface.[24]

The remoteness of the parks also assured, by and large, that they had little economic value, which dissipated industrial resistance to their establishment. Indeed, Congress regularly sought and received assurances that proposed parklands were "worthless,"[25] and some places that did have important commercial value—such as the coastal redwoods of California—were kept out of the system for more than half a century.[26] Only rarely did conflict become bitter in the old days, as when San Francisco and the Sierra Club battled over the damming of Hetch Hetchy Valley in Yosemite Park for municipal water supply.[27] In 1913 the city won and the Sierra Club still bears scars from that fight, but Hetch Hetchy was an exceptional case. By the time major battles began to be fought over industrialization versus preservation, as in the struggle to keep dams out of Yellowstone in the 1920s,[28] the national parks were already a solidly entrenched feature of American life.[29]

The happy convergence of many disparate interests permitted Congress and the public to sustain the contradictory, but compatible, beliefs that permitted a park system to flourish: on one side a repugnance at the seemingly boundless materialism that infused American life, a spiritual attachment to untrammeled nature, and a self-congratulatory attitude toward preservation of nature's bounty; and on the other a commitment to economic progress wherever it could be exacted, nationalistic pride, and the practical use of nature as a commodity supportive of tourism and commercial recreation.

For a good many years, this fragile ideological coalition held together with only modest conflict. The preservationists (as they are now called), who always comprised the most active and interested constituency in favor of national parks, had little to complain about. The parks were there, but they were so little used and so little developed—Congress was always grudging with appropriations: "Not one cent for scenery" was its long-standing motto[30]—that those who wanted to maintain the parks as they were, both for their own use and as a symbol of man's appropriate relationship to nature, had what they wanted.

The professional park managers, organized as the National Park Service in 1916, also found circumstances generally to their liking. Like all bureaucrats they had certain imperial ambitions. But the park system was steadily growing, and that was satisfying. Some of their gains were made at the expense of the national forests, housed in another federal department, and while interbureau infighting was at times intense, the general public was indifferent to such matters.[31] Moreover, in its early years, and particularly before the full blossoming of the automobile era, the Park Service was able to take an actively promotional posture, encouraging increasing tourism, road building, and hotel development without losing the support of its preservationist constituency.[32] It was then in everyone's interest to create greater public support for the parks. If more people came to the national parks, more people would approve the establishment of new parks and would approve funding for management needed to protect and preserve them. Even the most ardent wilderness advocate complained little about the Park Service as a promotional agency.

The adverse effects tourism might have were long viewed as trivial.

The tourists who came to the parks in the early days were in general not much different from those who come today.[33] They arrived in carriages, slept in hotels, and spent a good deal of their time sitting on verandas. But of course they came in much smaller numbers, their impact on the resources was much less, and, despite the comforts they provided themselves, the setting in which they lived in the parks was fairly primitive and marked a sharp contrast with life at home. A visit to a national park was still an adventure, quite unlike any ordinary vacation. The alliance of preservationists (whose interest in parks was essentially symbolic and spiritual) and vacationers (to whom the parks were a commodity for recreational use) was not threatened by the low intensity use the parks received for many decades. The contradiction Congress had enacted into law in the 1916 general management act, ordering the National Park Service at once to promote use and to conserve the resources so as to leave them unimpaired, was actually a workable mandate.[34]

The recreation explosion of recent years has unraveled that alliance and brought to the fore questions we have not previously had to answer: For whom and for what are the parks most important? Which of the faithful national park constituencies will have to be disappointed so that the parks can serve their "true" purpose? The adverse impact on natural resources generated by increased numbers is only the most visible sign of a cleavage that goes much deeper. The preservationist constituency is disturbed not only—and not even most importantly—by the physical deterioration of the parks, but by a sense that the style of modern tourism is depriving the parks of their central symbolism, their message about the relationship between man and nature, and man and industrial society.

When the tourist of an earlier time came to the parks he inevitably left the city far behind him. He may not have been a backpacker or a mountain climber, but he was genuinely immersed in a natural setting. He may only have strolled around the area near his hotel, but he was in a place where the sound of birds ruled rather than the sound of motors, where the urban crowds gave way to rural densities, and where planned entertainments disappeared in favor of a place with nothing to do but what the visitor discovered for himself.

Tourism in the parks today, by contrast, is often little more than an extension of the city and its life-style transposed onto a scenic background. At its extreme, in Yosemite Valley or at the South Rim of Grand Canyon, for example, one finds all the artifacts of urban life: traffic jams, long lines waiting in restaurants, supermarkets, taverns, fashionable shops, night life, prepared entertainments, and the unending drone of motors.[35] The recreational vehicle user comes incased in a rolling version of his home, complete with television to amuse himself when the scenery ceases to engage him. The snowmobiler brings speed and power, Detroit transplanted, imposing the city's pace in the remotest backcountry.

The modern concessioner, more and more a national recreation conglomerate corporation, has often displaced the local innkeeper who adapted to a limited and seasonal business. There are modernized units identical to conventional motels, air conditioning, packaged foods, business conventions, and efforts to bring year-round commercial tourism to places where previously silent, languid winters began with the first snowfall.[36]

All these changes have made the preservationist, to whom the park is essentially a symbol of nature and *its* pace and power, an adversary of the conventional tourist. The clearest evidence that the preservationist and the tourist are not simply fighting over the destruction of resources or the allocation of a limited resource that each wishes to use in different, and conflicting, ways, but are rather at odds over the symbolism of the parks, is revealed by the battles that they fight. One such recent controversy has arisen over the use of motors on concessioner-run boat trips down the Colorado River in Grand Canyon.[37] In fact, motorized boats don't measurably affect the Canyon ecosystem, nor do they significantly intrude upon those who want to go down the river in oar-powered boats. Reduced to essentials, the preservationist claim is simply that motors don't belong in this remote and wild place; that they betray the idea of man immersed in nature and bring industrialization to a place whose meaning inheres in its isolation from, and contrast to, life in society.

Much the same observation may be made about the intense controversy over highly developed places like Yosemite Valley. Many of those who are most opposed to the claimed over-development of the valley do not themselves use it much. Wilderness lovers go into the wilderness, and Yosemite, like most national parks, has an abundance of undeveloped wilderness. What offends is not the unavailability of the valley as wild country, but the meaning national parks come to have when they are represented by places like Yosemite City, as the valley has been unkindly called.

What's wrong with the parks, says Edward Abbey—one of the most prominent contemporary spokesmen for the preservationist position—is that they have been too much given over to the clientele of "industrial tourism," people who visit from their cars and whose three standard questions are: "Where's the john? How long's it take to see this place? and Where's the Coke machine?"[38] Perhaps serving vacationers who have questions like these on their minds would require the construction of some additional roads and the installation of a few more Coke machines, but those intrusions need hardly interfere with Abbey's own recreational preferences, particularly in the vast Utah parks he most admires. His complaint is of quite a different kind. Industrial tourism debases the significance that national parks have for him, and he is troubled to see people using the parks as they use Disneyland, simply as places to be entertained while they are on vacation.

Traditional approaches to conflicting uses in the parks are not responsive to the issue that really divides the preservationist and the tourist. It will not do simply to separate incompatible uses, or to mitigate the damage done by the most resource-consuming visitors. For the preservationist is at least as much interested in changing the attitudes of other park users as in changing their activities. And he is as much concerned about what others do in places remote from him as when they are vying for the same space he wants to occupy. The preservationist is like the patriot who objects when someone tramples on the American flag. It is not the physical act that offends, but the symbolic act. Nor is the offense mitigated if the trampler points out that the flag belongs to him, or that flag trampling is simply a matter of taste, no different from flag waving.

The preservationist is not an elitist who wants to exclude others, notwithstanding popular opinion to the contrary; he is a moralist who wants to convert them. He is concerned about what other people do in the parks not because he is unaware of the diversity of taste in the society, but because he views certain kinds of activity as calculated to undermine the attitudes he believes the parks can, and should, encourage. He sees mountain climbing as promoting self-reliance, for example, whereas "climbing" in an electrified tramway is perceived as a passive and dependent activity. He finds a park full of planned entertainments and standardized activities a deterrent to independence, whereas an undeveloped park leaves the visitor to set his own agenda and learn how to amuse himself. He associates the motorcyclist roaring across the desert with aspirations to power and domination, while the fly-fisherman is engaged in reducing his technological advantage in order to immerse himself in the natural system and reach out for what lessons it has to offer him. The validity of these distinctions is not self-evident, and I shall have a good deal more to say about them in the following chapters. They are, however, what lies at the heart of the preservationist position.

The preservationist does not condemn the activities he would like to exclude from the park. He considers them perfectly legitimate and appropriate—if not admirable—and believes that opportunities for conventional tourism are amply provided elsewhere: at resorts and amusement parks, on private lands, and on a very considerable portion of the public domain too. He only urges a recognition that the parks have a distinctive function to perform that is separate from the service of conventional tourism, and that they should be managed explicitly to present that function to the public as their principal goal, separate from whatever conventional tourist services they may also have to provide.

In urging that the national parks be devoted to affirming the symbolic meaning he attaches to them, the preservationist makes a very important assumption, routinely indulged but hardly ever explicit. The assumption is that the values he imputes to the parks (independence, self-reliance, self-restraint) are extremely widely shared by the American public. Though he knows that he is a member of a minority, he believes he speaks for values that are majoritarian. He is, in fact, a prophet for a kind of secular religion. You would like to emulate the pioneer explorers, he says to the public; you would like independently to raft down the wild Colorado as John Wesley Powell did a century ago.[39] You would like to go it alone in the mountain wilderness as John Muir did. Indeed that is why you are stirred by the images of the great national parks and why you support the establishment of public wilderness. But you are vulnerable; you allow entrepreneurs to coddle you and manage you. And you are fearful; you are afraid to get out of your recreational vehicle or your car and plunge into the woods on your own. Moreover you want to deceive yourself; you would like to believe that you are striking out into the wilderness, but you insist that the wilderness be tamed before you enter it. So, says the secular prophet, follow me and I will show you how to become the sort of person you really want to be. Put aside for a while the plastic alligators of the amusement park, and I will show you that nature, taken on its own terms, has something to say that you will be glad to hear. This is the essence of the preservationist message.

2

An Ideal in Search of Itself

The early preservationists and park advocates assumed, without ever explaining, that personal engagement with nature could build in the individual those qualities of character that the existence of the parks symbolized for us collectively. Perhaps the point was made most explicitly by the celebrated wilderness pioneer Aldo Leopold in his essay, "Wildlife in American Culture." "No one can weigh or measure culture," Leopold observed.

> Suffice it to say that by common consent of thinking people, there are cultural values in the sports, customs and experiences that renew contacts with wild things. . . . For example, a boy scout has tanned a coonskin cap, and goes Daniel-Booning in the willow thicket below the tracks. He is reenacting American history. . . . Again, a farmer boy arrives in the schoolroom reeking of muskrat; he has tended his traps before breakfast. He is reenacting the romance of the fur trade.[1]

Certainly it would seem eccentric to hold national parks simply so that people could go muskrat trapping. Like Aldo Leopold, John Muir and most other early park supporters had an idea in their minds about the importance to people of encounters with nature, but they seemed at a loss when it came to formulating their intuitions into any coherent recreational plan. To a substantial extent the presumption seems to have been that if only people would come into the parks, as John Muir put it, they would find "everything here is marching to music, and the harmonies are all so simple and young they are easily apprehended by those who will keep still and listen and look. . . ."[2]

But it wasn't simple at all, as Muir himself soon realized. Many came and looked, but they didn't see what he had seen, just as they listened without hearing what he had heard. National park admirers have frequently ignored the fact that nature has commended itself to people in very different ways at different times. The awesome grandeur of the parks has at times been thought fearsome rather than beautiful. It is perfectly possible to conceive of wilderness as something to be conquered rather than worshipped; people can, and have, shunned rather than climbed mountains. And it is quite as possible to respond to parks as pleasant sites for picnics and hotel resorts as to view them as fragile museums of nature or history.

Aside from scattered hints here and there, there is little serious or sustained writing to which we might turn for guidance in seeking to understand how those who conceived of parks as culturally important recreational resources meant them to be used. There is, however, at least one document that seeks explicitly to address itself to this question, a report entitled "The Yosemite Valley and the Mariposa Big Trees," written in 1865 by Frederick Law Olmsted.[3] Olmsted is not a name that leaps immediately to mind when one thinks of the national parks. He was of course America's premier landscape architect, and though he was a man of many remarkable accomplishments—including the authorship of a fine series of books on the pre–Civil War South, leadership in the United States Sanitary Commission which was the predecessor to the Red Cross, and innovative work in the design of suburban communities—he is known to most Americans only as the designer of Central Park in New York.

For a brief period, however, during 1864 and 1865, Olmsted left New York to become the manager of the troubled Mariposa mining properties in northern California. While there is no conclusive evidence, it is highly likely that he was one of a small band of Californians who urged the federal government to preserve Yosemite Valley and the Mariposa grove of giant sequoias from settlement and destruction.[4] Olmsted was appointed the first chairman of the board of commissioners that California established to manage the Yosemite Park; and during his brief chairmanship he wrote a report that was intended as a basis for future management. In it he also set out to explain why it was desirable to have a place like Yosemite as a public park, and in those observations lie the report's great interest.

Olmsted read his report to his fellow commissioners in August, 1865, but it was not published, and it then simply disappeared. It has been suggested that the report was suppressed by those in the California Geological Survey who feared that Olmsted's plan for Yosemite might create competition for legislative appropriations. Whatever the case, it was not until nearly ninety years later, in 1952, that diligent searching by Laura Wood Roper, Olmsted's biographer, turned up a virtually complete copy in the still-extant Olmsted firm's office in Brookline, Massachusetts.[5] Roper published the report in the magazine *Landscape Architecture,* where it remains largely unknown, though in it, as she justly remarks, "Olmsted formulated a philosophic base for the creation of state and national parks."[6]

The failure of Olmsted's report to command modern attention is less surprising than might at first appear.[7] Unlike much popular nature writing, the report lacks rapturous descriptions of self-discovery, and it is marred by a certain archaic nineteenth-century style of expression. Olmsted talks about the advance of civilization and speaks of "scientific facts," among which he numbers mental disabilities like softening of the brain and melancholy.[8] Some effort is required to penetrate these passages, but it is well worth making.

Olmsted begins at the beginning. The park was established for the preservation of its scenery. He does not, however, treat this as a self-justifying observation. The question is why government should take upon itself the burden of scenic preservation. His answer at one level is largely descriptive. Striking scenery has a capacity to stimulate powerful, searching responses in people. "Few persons can see such scenery as that of the Yosemite," he notes, "and not be impressed by it in some slight degree. All not alike, all not perhaps consciously, . . . but there can be no doubt that all have this susceptibility, though with some it is much more dull and confused than with others."[9] He does not claim to be making some universally true claim, good for all time, but certainly it was a claim that was true enough for his own time, and for ours. As Olmsted observed, Yosemite had become a popular subject for artists and photographers, and their widely reproduced works had induced a great interest in, and admiration for, the place. Moreover, in the Old World, it had long been a tradition to reserve the choicest natural scenes in the country for the use of the rich and powerful. Apparently people able to do whatever they wanted found great satisfaction could be elicited from engagement with striking scenery.

At this point, Olmsted offers his distinctive hypothesis—the basis of his prescription for the national parks. In most of our activities we are busy accomplishing things to satisfy the demands and expectations of other people, and dealing with petty details that are uninteresting in themselves and only engage our attention because they are a means to some other goal we are trying to reach. Olmsted does not suggest that gainful activity is a bad thing by any means; only that it offers no opportunity for the mind to disengage from getting tasks done, and to engage instead on thoughts removed from the confinement of duty and achievement. He calls this the invocation of the contemplative faculty.

For Olmsted the preservation of scenery is justified precisely because it provides a stimulus to engage the contemplative faculty. "In the interest which natural scenery inspires . . . the attention is aroused and the mind occupied without purpose, without a continuation of the common process of relating the present action, thought or perception to some future end. There is little else that has this quality so purely."[10]

Olmsted does not purport to explain why scenery has this effect on us, though doubtless the modern attraction to the idea of God-in-nature is a plausible explanation. He is content to observe that there is something that moves us to appreciate natural beauty and to be moved by it, and "intimately and mysteriously" to engage "the moral perceptions and intuitions."[11] He recognized that not everyone responds in this way, thus anticipating the objection that nature parks established for their scenery would not likely be as popular as amusement parks. But he attributed this to a lack of cultivation. It is unquestionably true, but it is not inevitable, he said, "that excessive devotion to sordid interests," to the constant and degrading work upon which most people are engaged, dulls the aesthetic and contemplative faculties.[12] It is precisely to give the ordinary citizen an opportunity to exercise and educate the contemplative faculty that establishment of nature parks as public places is "justified and enforced as a political duty."[13]

No one, he thought, was more relentlessly tied to unreflective activity than the ordinary working citizen. The worker spends his life in almost constant labor, and he has done so traditionally because the ruling classes of the Old World had nothing but contempt for him. They thought "the large mass of all human communities should spend their lives in almost constant labor and that the power of enjoying beauty either of nature or art in any high degree, require[d] a cultivation of certain faculties, which [are] impossible to these humble toilers."[14] Olmsted rejects this belief categorically. Behind his rather archaic vocabulary, and his psuedoscientific proofs, lies a prescription for parks as an important institution in a society unwilling to write off the ordinary citizen as an automaton.

Olmsted, as a practical man, set out a number of specific suggestions for the management of parks. He had an idea about the "thing" that should be made available to the public as a park, just as the curator has an idea of the collection to be presented in a museum.

The first point, he said, is to keep in mind that the park was reserved because of its scenery, and therefore the first task

> is the preservation and maintenance as exactly as is possible of the natural scenery; the restriction, that is to say, within the narrowest limits consistent with the necessary accommodation of visitors, of all constructions markedly inharmonious with the scenery or which would unnecessarily obscure, distort or detract from the dignity of the scenery.[15]

To read this formula in isolation is to have the impression that Olmsted was advocating a pure wilderness status for the parks or that he was interested only in an aesthetic or visual experience, but plainly this is not at all what he had in mind. His principal goal in seeking preservation of the scenery was to assure that there would be no distractions to impede an independent and personal response to experience. Olmsted did not have an ideological opposition to the presence of any particular structure, such as roads or hotels in the park, for, as we shall see, he found such developments perfectly acceptable. His concern was with the installation of facilities or entertainments where "care for the opinion of others"[16] might dominate, or where prepared activities would occupy the visitor without engaging him.

Thus, for example, Olmsted would have found the modern ski resort an anomaly in the parks, not because it intrudes upon the scenery, or impairs the indigenous ecosystem, or because of the skiing itself, but because of the crowding, commercialism, obtrusive social pressures, and the inducements to participate in entertainments planned and structured by others.

While he did not spell out his management theory in detail in the Yosemite report, he returned to the problem twenty years later in a report for a state park at Niagara Falls. Niagara had been the most popular tourist attraction in America during the later nineteenth century, but all the land had been sold into private ownership and commercial enterprises had taken over. Tourists were importuned and harassed, led around like trained animals and hurried from one "scenic site" to the next.[17]

As early as 1869, Olmsted began a campaign to establish a public park around Niagara Falls, and to combat the desecration of the area that had taken place.[18] The park was finally established in 1886; in 1879 Olmsted prepared a study proposing a management scheme for the Niagara Park,[19] and eight years later he drew up a detailed planning report.[20]

The Niagara report contains a passage almost identical to that quoted earlier from the Yosemite work, asserting that nothing of an artificial character should be allowed to interfere with the visitor's response to the scenery. But in the Niagara report,

Olmsted set out his views about park management in much more detail. Again, he made clear that a wilderness park need not be established. It would be quite appropriate to provide, near the entrance, toilets, shelters, picnic facilities, and the like. He also recommended the construction of walkways, as well as restorative efforts to combat erosion and revegetate barren areas.

He opposed fancy landscaping, however, because it is calculated to draw off and dissipate regard for natural scenery. For the same reasons he opposed a plan to build a fine restaurant on Goat Island, a wild place just above Niagara Falls. Neither, he said, ought sculpture or monuments to be placed within the park, worthy as they are.

Probably the most revealing expression of Olmsted's approach was his opposition to a proposal to permit people to see the falls without having to leave their carriages. This was not an obvious issue for him, for in the Yosemite report he had advocated the construction of a carriage road in the valley. But Yosemite, at that time, was a very remote place, with few visitors and difficult access. Niagara was entirely different, and Olmsted's response—based on different circumstances—tells a great deal about his conception of a rewarding park experience.

He began with the observation that as many as ten thousand people a day visited Niagara, and that to permit the scenic grandeur of the place to engage the visitor it was necessary to see the falls at length and at leisure. If the scenic viewing areas were designed to accommodate large numbers of carriages, it would "interpose an urban, artificial element plainly in conflict with the purpose for which the Reservation has been made." The purpose of the park was to encourage people to experience Niagara "in an absorbed and contemplative way." A profusion of carriages, with crowds of people, would intrude upon the opportunity for an independent experience.

He sought to restore the setting of an earlier Niagara, where

> a visit to the Falls was a series of expeditions, and in each expedition hours were occupied in wandering slowly among the trees, going from place to place, with many intervals of rest. . . . There was not only a

much greater degree of enjoyment, there was a different kind of enjoyment. . . . People were then loath to leave the place; many lingered on from day to day . . . revisiting ground they had gone over before, turning and returning.[21]

It is striking to see how far removed Olmsted's views are from the sterility of current battles over riding versus walking, or wilderness versus development. Olmsted believed that the essence of the park is not determined by the details of the visitors' activities, by whether they see the park from a sitting rather than a standing position, or sleep in a tent rather than a hotel bed. His attention was focused on the attitude that the visitor brought to the park, and upon the atmosphere that park managers provided for the visitor. He thought it perfectly possible to have an appropriate park experience using a vehicle in a remote enough place; just as he would, without doubt, have condemned the relentless backpacker whose principal concern is to prove that he can "do" so many miles a day, or climb more peaks than any of his predecessors. His goal was to get the visitor outside the usual influences where his agenda was preset, and to leave him on his own, to react distinctively in his own way and at his own pace.

To understand Olmsted's views it is essential to keep in mind that he was a republican idealist. He held, that is to say, to what we generally call democratic values. He believed in the possibility of a nation where every individual counted for something and could explore and act upon his own potential capacities. He feared, and he condemned, the nation of unquestioning, mute, and passive followers. The destruction of Niagara's scenery appalled him, not simply because the place was ugly, but because old Niagara was a symbol and a means for the visitor freely to respond to his experience. The trouble with the new Niagara was that it had returned, with its leading and hurrying of visitors and with its commercial entertainments, in the guise of free enterprise, to the same contemptuous disregard of the individuality of the visitor that had characterized the aristocratic, condescending spirit of Europe.

Olmsted was criticized on the ground that his plan for Niagara constituted an attack upon a place that was—for all its tawdry development—extraordinarily popular. The charge was, as

Olmsted rephrased it, that "whatever has been done to the injury of the scenery has been done . . . with the motive of profit, and the profit realized is the public's verdict of acquittal."[22]

He, of course, conceded Niagara's popularity, but it was his conviction that the best use of highly scenic areas was not to serve popular taste but to elevate it. The new Niagara was a modern version of precisely what he had condemned in the Yosemite report: the belief of the governing classes of Europe that the masses were incapable of cultivation. Hence, they had thought "so far as the recreation of the masses receives attention from their rulers, to provide artificial pleasures for them, such as theatres, parades, and promenades where they will be amused by the equipages of the rich and the animation of the crowds."[23] "The great body of visitors to Niagara come as strangers. Their movements are necessarily controlled by the arrangements made for them. They take what is offered, and pay what is required with little exercise of choice."[24]

The commercialized Niagara was enjoyable, it provided a service for the leisure time that citizens had to spend. Olmsted's Niagara plan called for some sacrifice of that service in order to provide a place designed to engage the contemplative faculty and to encourage the visitor to set his own agenda. He believed these were opportunities that citizens of a democratic society ought to want to provide themselves.

Olmsted's distinctive conception of a park is not easily captured in a phrase. He repeatedly uses the word "contemplative," but plainly it is not an intellectual experience he has in mind. He also talks about "cultivation" and "refinement," faintly archaic terms, that are probably nearest to our notion of the conscious development of aesthetic appreciation. Though he speaks principally of the visual experience of scenic inspiration—understandably enough in light of his professional work as a landscape architect—his Yosemite report also contains approving references to hunting and mountaineering. And there is a strong element in his writing of republican idealism, a distaste for the mass man unreflectively doing what he is told to do and thinking what he is told to think.

Of course Olmsted was himself a man of the nineteenth century, and his writing reveals a confident belief, characteristic of the time, in the progress of the human spirit. The attitude he evinces is reminiscent of the famous passage in Ralph Waldo Emerson's essay, "Nature":

> Adam called his house, heaven and earth; Ceasar called his house, Rome; you perhaps call yours, a cobbler's trade. . . . Yet . . . your dominion is as great as theirs, though without fine names. Build therefore your own world, As fast as you conform your life to the pure idea in your mind, that will unfold its great proportions.[25]

Olmsted's dedication to a spirit of independence also echoes Emerson. "The spirit of the American freeman is already suspected to be timid, imitative, tame," Emerson wrote in "The American Scholar."[26] Indeed, Olmsted's views draw on a pastoral, moral, and aesthetic tradition with even deeper roots.[27] The distinctiveness of his contribution lies in the application of these ideas to the public institution of a nature park, and therein lie some puzzling questions. What special activities and attitudes, for example, would be called for on the part of visitors to such parks; and how does one deal with the claim that as public facilities parks also have a responsibility to meet the demands of conventional tourism? Olmsted's work only hints at answers to such questions.

3

The Ideal in Practice

An extensive and largely ignored body of literature—produced not by scholars, but by the participants themselves—captures the essence of the reflective, independent qualities Olmsted sought to describe as the ideal for recreation in the national parks.[1] With rare exceptions, these writings have been treated as popular entertainment intended for an audience of fellow sportsmen. They deserve wider and more serious attention.

Probably no recreation has produced a larger body of books and articles than fishing.[2] On first consideration, the point seems obvious enough: People go fishing in order to catch fish. Yet the single theme that dominates the fishing literature is a disavowal of precisely this proposition. Arnold Gingrich, a well-known writer on the subject, opens his book *The Joys of Trout* with the recollection that "if a careful count were kept, it would show that over the last five years my evenings have been just a little more often fishless than not." Yet, he adds, "since I never keep the fish I catch anyway, a realist might well ask what difference it makes."[3] That is the question to which scores of fishing books have addressed themselves.

Certainly it would be misleading to suggest that catching fish is a matter of indifference to the serious fisherman. What is clear, though, is that fishing at its best is not *about* catching fish. Roderick Haig-Brown, a celebrated fly-fishing writer, captured the spirit of the literature when he wrote: "I do not fish for fish to eat . . . I do fish to catch fish . . . at least that is an idea not too far from the back of my mind while I am fishing; but I have fished through fishless days that I remember happily and without regret. . . ."[4] Albert Miller, who writes under the name Sparse Grey Hackle, picks up the same verbal formulation in the title to his best known book, *Fishless Days, Angling Nights.* Miller's book opens with the statement, "Fortunately, I learned long ago that although fish do make a difference—*the* difference—in angling, catching them does not"; the secret of fishing is to be "content to not-catch fish in the most skillful and refined manner. . . ."[5] It is no coincidence that Miller adopts one of Olmsted's favorite nineteenth century words, refinement. Fishing is most satisfying, not when it results in accomplishment of a set task, but in refining us.

In the greatest of all fishing books, Walton and Cotton's *The Compleat Angler,* the narrator Piscator replies to those who pity the ardent fisherman, comparing him unfavorably to purposeful, serious men of affairs.

> Men who are taken to be grave . . . money-getting men, men that spend all their time, first in getting it, and next in anxious care to keep it; . . . we Anglers pity them perfectly . . . and stand in no need to borrow their thought to think ourselves so happy.[6]

If fishing were only the getting of fish, Piscator says, it would be nothing but an outdoor version of what "these poor-rich-men" do. And when his companion notes in frustration that he has followed Piscator for two hours and not even seen a fish stir, he is told that he has not yet learned what angling is all about. "There is more pleasure in hunting the hare than in eating her. . . . As well content no prize to take / As use of taken prize to make."[7]

The subtitle of *The Compleat Angler* is *The Contemplative Man's Recreation,* and here again the verbal similarity with Olmsted's definition of the park, as a place designed to stir the contemplative faculty, is revealing. Angling is an art, and fishing is simply the raw material of that art, whereby the mind is engaged; a good angler must bring to his recreation "an inquiring, searching, observing, wit."[8] One of the most famous passages in Walton and Cotton's book compares angling to mathematics: "It can never be fully learned . . . an art worthy the knowledge and practice of a wise man."[9] In the charming autobiographical story, "A River Runs Through It," Norman MacLean says "it is not fly fishing if you are not looking for answers to questions."[10] And Roderick Haig-Brown speaks of fly-fishing as an activity calculated to evoke "the subtle and difficult things:"[11]

> I can lie for hours at a time and watch the flow of a little stream . . . the secret vagaries of current are clearly revealed here. . . . A fold or break of current, a burst of bubbles or the ripple of a stone . . . releases in me a flood of satisfaction that must, I think, be akin to that which a philosopher feels as his mind is opened to a profound truth. I feel larger, and better and stronger for it in ways that have nothing to do with any common gain in practical knowledge.[12]

These descriptions raise a question to which the fishing literature gives no direct answer. Is it simply the setting, the fascinating stream or the grand scenery? Or is there something about the activity itself essential to production of the profound satisfaction he describes? Neither the setting nor the activity *in itself* seems to be decisive; rather, it is the presence of something capable of engaging, rather than merely occupying, the individual—a stimulus for intensity of experience, for the full involvement of the senses and the mind.

The setting may be important because of its complexity or its unfamiliarity. A trout in a trout stream is more provocative than a trout in a fishbowl; an undeveloped forest is more likely to engage our concentration than the cornfield we see every day. Of course there are no absolutes here. To a scientist, a common cornfield may be endlessly fascinating and puzzling, and to the artistic eye the most common events may be dazzling. For Proust nothing more was required than the routine of a mother's goodnight kiss, the tedious salons of Paris, and the daily events of a banal seaside resort. Most of us are not so discerning; for us setting counts.

The activity counts too. Fishing for the wily trout in its natural habitat forces us to be attentive to the smallest detail in a way that driving by at a high speed, or a casual walk, may not. It's not only what we do, but what we refrain from doing. The installation of snack stands and souvenir shops at Niagara were a distraction calculated to divert the visitor from intense concentration upon anything, while the majestic grandeur of the falls has a capacity to focus our attention. The presence of concessioners offering preplanned pony or boat rides can be an impediment to intensity of experience, diverting us from coming at the experience in our own way and at our own pace.

The facilities we provide for ourselves also affect our responses. To drive through the desert in an air-conditioned car is an insulating experience. The increasingly popular recreation of backpacking offers a revealing counterexample.[13] Hiking with a pack on one's back appears superficially to be a strangely unappealing activity. The hiker, vulnerable to insects and bad weather, carries a heavy load over rough terrain, only to end up in the most primitive sort of shelter, where he or she eats basic foods prepared in the simplest fashion. Certainly there are often attractive rewards, such as a beautiful alpine lake with especially good fishing. But these are not sufficient explanations for such extraordinary exertions, for there are few places indeed that could not be easily made more accessible, and by much more comfortable means.

To the uninitiated backpacker a day in the woods can be, and often is, an experience of unrelieved misery. The pack is overloaded; tender feet stumble and are blistered. It is alternately too hot or too cold. The backpacker has the wrong gear for the weather or has packed it in the wrong place; the tent attracts every gust of wind and rivulet of water. The fire won't start, or the stove fails just when it's needed. And the turns that seemed so clear on the map have now become utterly confusing.

Such experiences, familiar in one form or another to all beginners, are truly unforgiving; and when things go wrong, they do so in cascading fashion. Yet others camping nearby suffer no such miseries. Though their packs are lighter, they have an endless supply of exactly the things that are needed. Their tents go up quickly, they have solved the mystery of wet wood, and they sit under a deceptively simple rain shelter, eating their dinner in serene comfort. What is more, they are having a good time. The woods, for the beginner an endless succession of indistinguishable trees apparently designed to bewilder the hapless walker, conceal a patch of berries or an edible mushroom. Nearby, but unseen, are beautiful grazing deer or, overhead, a soaring eagle.

With time, patience, and effort one recognizes that these things are available to everyone; it is possible to get in control of the experience, to make it our own. The pack lightens as tricks are learned: how to substitute and how to improvise quickly, out of available materials, the things previously lugged. The more known, the less needed. Everything put in the head lessens what has to be carried on the shoulders. The sense of frustration falls away and with it the fear that things will break down. One knows how to adapt. The pleasure of adaptation is considerable in itself because it is liberating.

Nor is it merely a lifting of burdens. The backpacker, like the fisherman, discovers that the positive quality of the voyage is directly related to his or her own knowledge and resources. There is often a dramatic revelation that the woods are full of things to see—for those who know how to see them.

The kind of encounter that routinely takes place in the modern motorized vehicle, or in the managed, prepackaged resort, is calculated to diminish such intensity of experience. Nothing distinctive about us as individuals is crucial. The margin of error permitted is great enough to neutralize the importance of what we know. If we roar off in the wrong direction, we can easily roar back again, for none of our energy is expended. It isn't important to pay close attention to the weather; we are insulated from it. We need not notice a small spring; we are not at the margin where water counts. The opportunity for intensity of experience is drained away.

It is not that the motorized tourist or the visitor at a highly developed site must necessarily lose intensity, or that he is compelled to experience his surroundings at a remove, just as it is not inevitable that backpacking or fly-fishing will produce profound, individual responses. It is rather that the circumstances we impose on ourselves have the power to shape our experience.

The contrast between insulation and intensity is also demonstrated by the tools we use. Fishermen are probably more interested in equipment than are the devotees of any other leisure activity, and fishing books are full of endless discussion of flies, lines, rods, and leaders. Yet that interest is not at all directed to technological advance leading to increased efficiency in catching fish. Indeed, in one respect, it has exactly the opposite purpose: it is designed to maintain and even to increase the difficulty of success. At the same time, intricacy for its own sake is not sought. The goal is to raise to a maximum the importance of the participant's understanding, to play the game from the trout's point of view, so as to draw, as Haig-Brown puts it, upon "imagination, curiosity, bold experiment and intense observation."[14] This distinction between technology and technique is perhaps the most familiar common element in the recreational literature.

The hunting literature is very explicit in this respect though, like fishing, it at first seems wholly built around the conquest of a prey. One of the most provocative books ever written about that sport is the *Meditations on Hunting* of the Spanish philosopher José Ortega y Gasset.[15] Ortega's book was begun as a preface to another writer's conventional book about hunting, but it expanded into a full volume as he pondered the question, Why do we hunt? He was impressed by the fact that people have hunted over many centuries, and that the essence of the activity has not changed. A principal premise of the book is that rather than using every technological advantage available to him, the hunter has self-consciously neutralized his technological advantage in favor of the opportunity to develop what Ortega called technique:

> For hunting is not simply casting blows right and left in order to kill animals or to catch them. The hunt is a series of technical operations, and for an activity to become technical it has to matter that it works in one particular way and not in another. . . . It involves a complete set of ethics of the most distinguished design.[16]

To describe the hunting of animals as an ethical activity at first seems highly eccentric. Yet the recreation literature gives powerful support to Ortega's cryptic statement. The proposition that accomplishment is not of the essence is substantiated by a uniform view that the game gets better the more the player is able to intensify the experience. One practical application of this hypothesis is to disembarrass oneself of equipment whose purpose is simply to increase the ability to prevail.

The celebrated American wilderness advocate, Aldo Leopold, wrote about hunting in terms quite similar to those of the Spaniard Ortega. "There is," Leopold said, "a value in any experience that exercises those ethical restraints collectively called 'sportsmanship'. Our tools for the pursuit of wildlife improve faster than we do, and sportsmanship is a voluntary limitation in the use of those armaments."[17]

Leopold goes on to say something about hunting that is reminiscent of Olmsted's perception of recreation as a contrast to achievement. In the Yosemite report Olmsted not only spoke of accomplishment, but used the phrase "accomplishing something in the mind of another," that is, doing something because it wins the admiration of others. The fishing writers respond by observing that they are engaged in an activity that is judged only by the standard the fisherman sets for himself. And Leopold notes, "a peculiar virtue of wildlife ethics is that the hunter ordinarily has no gallery to applaud or disapprove of his conduct. Whatever his acts, they are dictated by his own conscience rather than by a mob of onlookers. It is difficult to exaggerate the importance of this fact."[18]

The attitudes associated with an activity may be more important than either the activity itself or its setting. To the extent that we infuse the parks with symbolic meaning by the way in which we use them, the symbolism attached to particular uses itself becomes a critical factor in the meaning that parks have for us. Consider, for example, the controversial question of off-road motorized vehicles (ORVs).[19] While ORVs have sometimes caused great and long-lasting damage, the vehicle itself is not the crucial factor in the controversy its use has created, for it is possible to imagine the lonely cyclist exploring the backcountry in quite the same fashion as the hiker or the horseman.[20]

Yet, in fact, the ORV has associated itself in our minds with a style of use that is quite at odds with Leopold's description of the ethical hunter, Olmsted's contemplative visitor, or Walton's pensive fisherman. The ORV has become a symbol of speed, power, and spectacle. The best-known ORV event on the public lands is the Barstow–Las Vegas motorcycle race that occurs on the California desert. Pictures of as many as three thousand cycles lined up to make the 150-mile crosscountry course have been widely published, both in books and on television.[21] This mass event, infamous for its destruction of the desert ecosystem, its rowdiness, and its vandalism, has become an emblem of the ORV. Commercial advertising has reinforced this picture, as publicity for off-road vehicles demonstrates: "Just put your gang on Suzuki's DS trail bikes. And head for the boonies. . . . Peaks or valleys, it's all the same to these rugged off-road machines. Tractoring up a hillside or going flat-out on a dry lake is no sweat."[22]

The descriptive literature provides a parallel image. In Lee Gutkind's book, *Bike Fever,* a day's expedition is reported as follows:

> The [motorcycle] bellowed as it bounced over the sage, and folded down the yellow grass on either side of the wheels. . . . He jetted off across the prairie for a while, breathing in the red dust that the wind and his wheels were kicking up. . . . He trampled the sagebrush . . . he had run into some "whoop-de-do" jumps—a series of brief hills, about 25 feet apart. He cranked on, climbed the hill, and disconnected from the ground. . . . Each time he hit the top of a hill, his wheels left the ground and his stomach ricocheted into his throat. . . .[23]

The picture here is all exhilaration and excitement—speed, danger, and domination. As a book entitled *The Snowmobiler's Companion* puts it,

> the snowmobile has brought back some of that edge-of-danger excitement, those feelings of man-against-the-elements adventure and man-over-machinery mastery that have been lost in every other form of modern transportation. . . . Why? To win. . . . To put on a spectacle. . . . To risk a life to the unending delight of hundreds of faces jammed up against the fences, mad for action, for crashes and beer.

Why? To prove that the machine is faster, the racer braver, better than the rest. To prove to whom? To Harry down the road. To yourself. To the faces at the fence.[24]

The ORV has become an extreme example of one kind of symbol, just as the motor-home recreational vehicle has of another—that of the passive visitor, unable to leave home and its comforts behind, sitting watching TV in the midst of the nation's most magnificent country. Other controversial uses—hang gliding, for example—emit a much less clear message, and to that extent engender much more ambivalent feelings. To some extent there is uneasiness because the activity seems a sort of spectacle of thrill seeking, rather like going over the falls in a barrel or riding a roller coaster. Conversely, the skills it requires, such as close attention to and understanding of complex wind patterns, make it seem rather like the activity of the hunter or fisherman who has minimized his tools and put himself as close to the margin of experience as possible.

These wide-ranging examples suggest an issue of subtlety and sophistication barely hinted at in Olmsted's writings. He asserted that activities removed from mere will to accomplishment and achievement in the eyes of others was important as a contrast to the values that so often dominate our daily lives. The fishing and hunting books clearly affirm that proposition. The cycling writings also speak to a kind of contrast—the passive twentieth-century citizen getting into active control of something and mastering it. While each seems to respond to similar longings, in practice they diverge sharply. The hunting and fishing writers are drawn to activities that transcend, without denying, the raw impulsion to exhibit power, win the game, pile up a score, and exercise dominion—treating the will to prevail as something natural, but at the same time dealing with it as something to be faced and measured, rather than yielded to.

Nowhere in the literature is this insight more explicit than in the rich stock of books on mountaineering.[25] There is a special intrigue in turning to this source, for among those who have comprised the national parks constituency over the years there is probably no recreation that has been more amply represented than mountain climbing. The Sierra Club, to take but one example, was for many years, in many ways, largely a mountaineering club; and John Muir, its patron saint, was, of course, John of the mountains.

It is impossible to read the climbing books without a certain mixture of attraction and repulsion. Particularly if one comes to them in the light of Olmsted's gentility, and his aesthetic sensibility, it is slightly shocking to read the tales of dogged determination, competitive striving to be first to the top, and unattractive infighting among members of climbing parties. The literature spans a wide spectrum from individual hiking to expedition climbing of the Mount Everest type. The latter is, obviously, quite a limited genre in terms of the numbers of people involved, but it has nonetheless been a primary source of published, and widely read, books. It has set the standard of style and rules of the game for those attracted to the mountains, just as Walton and Cotton or Haig-Brown have for fishermen.

What is one to make of these extraordinary books, with their reports of multimillion dollar expeditions, multitudes of hired porters, and diplomatic negotiations to assure primacy in reaching some remote summit? Thoreau said that only daring and insolent men climb mountains,[26] and one need not read very deeply in this literature to understand what he meant. Even the titles of the books are revealing. Among recent and popular publications, two of the best known are *Everest the Hard Way* (with the emphasis on *hard*),[27] and *In the Throne Room of the Mountain Gods*.[28] While the latter of these titles was probably sardonically chosen, the book being a rare effort to avoid the conventional glorifying style of the genre, it nonetheless conveys an accurate sense of what mountaineers think they are getting at—or getting to.

In many respects, mountain climbing books present a restatement of familiar themes. It is repeatedly observed that climbing at its best eschews the presence of an audience, and the longing is often expressed that "expeditions would go secretly and come back secretly, and no one would ever know."[29] The technique/technology distinction is sharply drawn, with much condemnation of the gadgetry that promotes success at the expense of the climber's opportunity to respond to the distinctive challenge each mountain presents.[30] There is understandable disdain for such astonishing decisions as the use of helicopters to negotiate the most difficult parts of Mount Everest, of which the famous English climber Chris Bonington said gently, it "seemed an unpleasant erosion of the climbing ethic."[31] More generally, the literature affirms the proposition that "climbing with a few classic tools that become extensions of the body is quite conducive to the sought-after feeling; using a plethora of gadgets is not."[32]

Likewise it is repeatedly observed that the essence of mountaineering is not reaching the summit, but the climb itself. "Reaching the summit of a mountain is not all it is cracked up to be," Galen Rowell says, "the summit is merely the curtain falling on a grand play."[33] Some years ago, the English alpinist Geoffrey Winthrop Young said, "in great mountaineering, the result, the reaching of a summit, is of minor importance . . . the whole merit of the climb depend[s] upon the way it was done, that is the method, behavior and mental attitude of the climbers. . . ."[34]

At the same time, there is a quality in mountaineering books of drive and competition, of a will to achievement, self-testing, and supremacy. Competitive drive is a quality far removed from what Olmsted was describing and from the attitude of America's greatest mountain explorer, John Muir. The struggle that is so central to most of this literature is, with a single exception—the night on Mount Shasta, recounted with great drama in *Steep Trails*—wholly absent in Muir's writing.[35] One of the lovely stories told about Muir is that after reading a magazine article in which a climber described his exciting perils in the ascent of Mount Tyndal, Muir remarked that the author "must have given himself a lot of trouble. When I climbed Mount Tyndal," he said, "I ran up and back before breakfast."[36]

At the heart of most writing about mountain climbing there is something very different from the experience of attunement that Muir and most other popular nature writers describe. At one level, it *is* the competitive striving that Olmsted sought to put aside, the "work hard, play hard" ethic associated with the ORV by which the standards and practices of the day-to-day world are imported whole into recreational activity. To this extent the climbing literature seems anomalous.

But there is another, and fascinating, element in these books. It is a picture of mountaineering as attractive to those who are strongly inclined to competition and striving, but serving as a means to come to terms with those intuitions in an activity whose traditions and style are calculated to transcend them. Galen Rowell's book, *In the Throne Room of the Mountain Gods,* contains numerous passages directed to just this point:

All of us by now were aware that the approach march was turning into a contest and that we were being judged in part by our pack weights and hiking times. . . . [M]y pack was frequently hefted by [others]. One would say, "Wow, that's light." . . . I'd like to be able to say that I wasn't bothered by these taunts. . . . Other things were more important to me. Or were they? One part of me longed to prove myself. . . . I, whether I admitted it to myself or not, was definitely competing when I matched my pace to that of the front-runners.[37]

In an entry in his diary, Rowell returned to this theme:

Most Western people, like dogs chasing their tails, devote their lives to a conscious pursuit of happiness. . . . Those of us hoping to climb K-2 have widened the circle of the chase. We are after a tangible goal—the summit of a mountain—which will function in our lives exactly as a material possession would, except that it will be nontransferable, theft-proof, and inflation-proof. Our society will register the achievement on an equal level with other, less abstract rewards of Western living. "I'd like you to meet Mr. Jones, the president of our local bank. And this is Mr. Dunham; he climbed the second highest mountain in the world."[38]

This, of course, is the same author who says that getting to the top is not the important thing, and that climbing is best when climbing alone or with a few quiet companions, not trying to follow someone else's standards for a climb. The impressive feature of Rowell's book is its rare openness, not only about the brutality of expedition climbing at its worst, but about the difficulty of achieving the sublime pleasures of a self-defining experience to which most such books are almost exclusively devoted.

The climbing experience at its best—"enjoyed purely for itself,"[39] as Rowell puts it, adopting almost the identical words Olmsted used in the Yosemite report—requires a detachment from the pressure of conventional expectations that is extremely difficult to achieve. The interest of climbing is not simply that it tends to attract those who feel these external pressures sharply, but that it induces the participant to confront this inner conflict rather than conceal it. Mountain climbing is a particularly interesting model because it draws together elements of skill development, tension between achievement and contemplation, independence, physical setting, and an established ethic. In an article entitled "Games Climbers Play,"[40] Lito Tejada-Flores notes that informal rules have evolved for various kinds of climbing experiences, set out as a series of negative injunctions: Don't use fixed ropes, belays, pitons, etc. The purpose of these rules is to build an ethical structure for the climbing game. "[T]hey are designed to conserve the climber's feeling of personal (moral) accomplishment against the meaninglessness of a success which represents merely technological victory."[41] Moreover, based on one's own level of skill and ability, each individual can select a kind of climbing game that is challenging for him. The idea is *not* that some games are better, harder, or more worthwhile in themselves than others, Tejada-Flores notes. Indeed, the very purpose of the game's structure is "to equalize such value connotations from game to game so that the climber who plays any of these games by its proper set of rules should have at least a similar feeling of personal accomplishment."[42]

At the same time, the climb is not simply a physical challenge or a series of dangerous moments. Its setting and pace provide an opportunity and incentive for intensity of experience beyond the physical. It is, the climber Doug Robinson suggests, "seeing the objects and actions of ordinary experience with greater intensity, penetrating them further, seeing their marvels and mysteries, their forms, moods, and motions . . . it amounts to bringing a fresh vision to the familiar things of the world."[43] A concentrated immersion in the natural scene, growing out of the pace of the climb and its demand for intense concentration, produces a special kind of observation. Here, for example is a description of a climb in Yosemite by Yvon Chouinard:

> Each individual crystal in the granite stood out in bold relief. The varied shape of the clouds. . . . For the first time we noticed tiny bugs that were all over the walls, so tiny they were barely noticeable. While belaying, I stared at one for 15 minutes, watching him move and admiring his brilliant red color.[44]

To be sure, not every climbing experience, or every climber, ascends either to such physical or mental peaks. Recent reports of a commercial enterprise devoted to getting beginners to the top of Mount Rainier, even if they have to be pulled up, make clear that no activity in itself has magic.[45] But mountaineering seems a particularly vivid example of the ideals and struggles with inner conflict that have fueled the recreational symbolism of the national parks.

The interlocking themes of the climbing literature—domination mediated by self-conscious restraint—are also powerfully reflected in the American literary tradition. Nowhere are they more fully realized than in Faulkner's "The Bear," the mythic hunting story of a yearly rendezvous with the great bear—symbol of the wilderness—"which they did not even intend to kill," not because it could not be vanquished but because the *mere* act of conquest would be merely an act of destruction.[46] The wilderness could be conquered, was being conquered, not by true hunters but by destroyers, "men with plows and axes who feared it because it was wilderness, men myriad and nameless even to one another," for whom wilderness had never "loomed and towered" in their dreams. The hunter's appointment with the bear is an inner rendezvous, a test of "the will and hardihood to endure and the humility and skill to survive,"[47] of men not yet tamed and not needful of taming the world around them.

A parallel theme runs through Hemingway's writing, even in the early "Big Two-Hearted River."[48] Everything in the previously described fishing literature is present there—the gentle day, the timelessness, the deep pleasures of getting intensely into the flow of the river, the unimportance to the fishing trip of catching fish. But the story obtains its power from the clearly felt but unstated fact that Nick Adams is not just whiling away a day on the river. He is exorcising a demon deep inside him.

The feeling of being at home and in harmony with things, the satisfying fatigue after a hard day of self-imposed labor, the pleasures of elemental truths intensely felt, the movement of the trout, the color of the grasshopper, the form of the landscape, the smell of food, are fully realized. But all this is overlain with an ominous sense of the pressures and perils in the world to which he will soon return. "He felt he had left everything behind, the need for thinking, the need to write, other needs. It was all back of him. . . . Nothing could touch him."[49] But these are not statements, they are questions. Sandwiched in the collection of stories entitled *In Our Time,* between two vivid descriptions of man's inhumanity to man, the final impression is of Nick's inevitable return to the conventional, and brutal, world outside. This is the literature of struggle.[50]

In Hemingway's late story, *The Old Man and the Sea,* the question of the hunt is posed in its starkest form.[51] Man strives for mastery and yet finds triumph only when he recognizes that he is not master. The desire to prevail is treated as natural: Santiago was born to be a fisherman just as the fish was born to be a fish.[52] But just as surely we know that victory alone is hollow; indeed, as has often been remarked in noting images of the crucifixion in the book, there can be victory *in* defeat where success is something other than conquest. The old man is beyond sentiment, as he is beyond proving himself to anyone, and this is what rescues the venture from meaningless sacrifice or wanton slaughter. It is the fisherman's ability to accept the inevitability of the struggle, without sentiment and without moralizing, that invests the venture with nobility. "Fish," he said, "I love you and respect you very much. But I will kill you dead before this day ends."[53]

From Olmsted to Faulkner and Hemingway by way of mountain climbers seems a tortuous route, but it is not nearly so indirect as first appearances suggest. The first step is detachment from conventional expectations and imposed obligation, for which the natural setting is a stimulus and a context. The sense of detachment that engagement with nature stimulates brings to the surface atavistic longings, while the "ethical" structure of activities like fishing and mountaineering constrains that atavism from becoming a mere will to conquer. The strong attraction of nature for denizens of modern industrial society draws its power from these elements. Engagement with nature provides an opportunity for detachment from the submissiveness, conformity, and mass behavior that dog us in our daily lives; it offers a chance to express distinctiveness and to explore our deeper longings. At the same time, the setting—by exposing us to the awesomeness of the natural world in the context of "ethical" recreation—moderates the urge to prevail without destroying the vitality that gives rise to it: to face what is wild in us and yet not revert to savagery.

From this perspective, what distinguishes a national park idea from a merely generalized interest in nature may be the special role that the nature park plays as an institution *within* a developed and industrialized society, in contrast to those traditions in which nature is offered as an alternative to society. The setting of the national park provides an opportunity for respite, contrast, contemplation, and affirmation of values for those who live most of their lives in the workaday world.

Unlike the pure pastoral tradition, the park does not proffer a utopian community of escape to a life of perfect harmony, forever free of conflict and besetting human passions.[54] Neither does it resemble what Henry Nash Smith, in his fine book *Virgin Land,* calls the myth of the West, an image of life beyond the frontier of civilization.[55] The failed western hero in American literature, as Smith makes clear, was an anarchic figure, a symbol of freedom beyond law and beyond constraint, modeled on an antithesis between nature and civilization. Conversely, the preservationist tradition in the national parks movement proposes no permanent escape from society to a utopian wilderness. Olmsted certainly was a civilized man, and much of his professional work was devoted to the design of urban parks for urban people. "We want," he said, "a ground to which people may easily go after their day's work is done . . . the greatest possible contrast with the streets and the shops and the rooms of the town. . . . We want, especially, the greatest possible contrast with the restraining and confining conditions of the town. . . ."[56] The same is true of the American nature writers. John Muir sought to build no communities in the mountains he tramped.[57] Just as Hemingway's fictional Nick Adams must come back from his idyllic fishing trip, so, characteristically, the modern wilderness pioneer, Bob Marshall, says in his Alaska journal: "In a week [I shall be back] in Seattle and the great thumping world. I should be living once more among the accumulated accomplishments of man. The world . . . cannot live on wilderness, except incidentally and sporadically."[58]

Engagement with nature as a prescription for man in society, rather than as a rejection of society, is nowhere more evident than in the work of Henry David Thoreau. Tameness and wildness are the terms Thoreau uses to express the tension between submissiveness and dominance that has emerged as a central motif in the preceding pages.

"Once or twice," Thoreau says in *Walden,* "while I lived at the pond, I found myself ranging the woods, like a half-starved hound, with a strange abandonment, seeking some kind of venison which I might devour, and no morsel could have been too savage for me. The wildest scenes had become unaccountably familiar."[59]

There is something primitive and frightening in these feelings, and yet something even more frightening in repressing them. When civilized attitudes tame us to the point that the instinct to prevail no longer weighs upon us, when we only think of animals as sides of beef to be eaten, we may do something worse than killing animals; we obliterate the problem of the kill from our consciousness. The hunter recognizes the problem because he is in touch with it; the ethical dilemma is still real for him because he knows the objects of his hunt face to face.[60] It is therefore not surprising to find Thoreau, though he himself ultimately abstained from hunting and fishing, saying that "perhaps the hunter is the greatest friend of the animals hunted, not excepting the Humane Society."[61] When Thoreau speaks of leaving the gun and fish pole behind, it is with a hope that we will, having struggled with the deepest forces in us, ultimately resolve the savage longing. He recognizes that the satisfaction of fishless days is not something easily or obviously come by, but is the product—at best—of a lifetime of reaching out for understanding. Those who came to fish at Walden during his residence, he says, commonly did not think they were lucky or well paid for their time unless they got a string of fish,

> though they had the opportunity of seeing the pond all the while. They might go there a thousand times before the sediment of fishing would sink to the bottom and leave their purpose pure; but no doubt such a clarifying process would be going on all the while.[62]

Thoreau's favorite word is wildness, and perhaps his most famous phrase "in wildness is the preservation of the world."[63] But plainly wildness does not mean the unthinking savage to Thoreau, as his revulsion at the primitivism he encountered in *The Maine Woods,*[64] or his uneasiness about the wholly uncultivated woodchopper he describes in *Walden,*[65] makes clear. Nor does it mean a world of untrammeled wilderness, as his attraction to agricultural pursuits demonstrates. Thoreau never left Concord society behind him, for he was always—both before and after Walden—a Concord man. He rather escaped the social values and conventions that dominated the town. He saw the people of Concord bored and boring, because they have been tamed.[66] And he sees in the woods around him a world which is characterized by nothing so much as its resistance to taming.

To be tamed is to be what someone else wants you to be, to be managed by their expectation of your behavior, to accept their agenda, to submit to their will, and to be dependent on their knowledge or largess. Dominance and submissiveness are only two versions of the same instinct. In "Walking," Thoreau is at his most explicit in setting out the philosophical thesis that underlies what he says elsewhere:

> I love even to see the domestic animals reassert their native rights—any evidence that they have not wholly lost their original wild habits and vigor; as when my neighbor's cow breaks out of her pasture. . . . I rejoice that horses and steers have to be broken before they can be made the slaves of men.[67]

Thoreau, unlike the "nature writers" with whom he is usually associated, conceives his response to nature in a form that is distinctively applicable to the situation of civilized society. We are at our best when we have not been tamed into the passivity of stock responses, of dependency, of insulation from intensity of experience.[68] To be willing to fish or climb without an audience; to be able to draw satisfaction from a walk in the woods, without calling on others for entertainment; to be content with a fishless day, demanding no string of fish to be counted and displayed: These are the characteristics of an individual who has "refined" wildness without taming it into the personality of the mass man. What the fisherman feels lying at the side of the brook watching the bubbles, or the mountain climber experiences as "purity of consciousness," are each versions of what psychologists describe in terms of personality as a "wonderful capacity to appreciate again and again, freshly and naively the basic goods of life, with awe, pleasure, wonder, and even ecstasy, however stale these experiences may have become to others."[69] Thoreau's writings—directed to his neighbors, living lives of "quiet desperation"—reveal the experience of one who pursues his own style, unencumbered by the preconceptions or expectations of others, finding the world, even in its most mundane elements, endlessly interesting because he approaches it intensely and searchingly.

The fundamental claim for what may be called reflective or contemplative recreation, then, is as an experimental test of an ethical proposition. Such recreation tests the will to dominate and the inclination to submissiveness, and repays their transcendence with profound gratification. Plainly such activities are not limited by any specific forms. They range from the purely contemplative wanderer in the woods who, like Thoreau or John Muir, has the capacity to detach himself from social convention and structured activity, to the agile climber arduously working his way to the meaning of the summit. Nor is the setting of nature an indispensable precondition. There is, for example, a strong commonality between the writings examined here and that of the Zen approach to sports. That literature too emphasizes intensity, skill development as an intermediate end, introspection, and—most significantly—a focus on the battle within. The classic work on the subject is Eugen Herrigel's *Zen in the Art of Archery,* and it parallels the nature literature quite closely.[70] Herrigel's work is devoted to the compelling proposition that "the art of archery means a profound and far-reaching contest of the archer with himself."[71] The author describes the culmination of his training as that moment when he finally understood the artless art of feeling "so secure in ourselves" that neither the score, nor the spectators, nor any external element remained important to him.[72]

While nature is not a uniquely suitable setting, it seems to have a peculiar power to stimulate us to reflectiveness by its awesomeness and grandeur, its complexity, the unfamiliarity of untrammeled ecosystems to urban residents, and the absence of distractions. The special additional claim for nature as a setting is that it not only promotes self-understanding, but also an understanding of the world in which we live. Our initial response to nature is often awe and wonderment: trees that have survived for millenia; a profusion of flowers in the seeming sterility of the desert; predator and prey living in equilibrium. These marvels are intriguing, but their appeal is not merely aesthetic. Nature is also a successful model of many things that human communities seek: continuity, stability and sustenance, adaptation, sustained productivity, diversity, and evolutionary change. The frequent observations that natural systems renew themselves without exhaustion of resources, that they thrive on tolerance for diversity, and they resist the arrogance of the conqueror all seem to give confirmation to the intuitions of the contemplative recreationist.

4

Making a Choice

Everything said up to this point implies that we can choose our recreation as freely as we choose our clothes. But there is a strong strain of contrary opinion that is rarely made explicit in debate over the national parks. Recreation fills needs created by the style of our daily lives, this view holds; and one need only know how someone works to know how he will play. The much-discussed problem of elitism arises from this perspective. For if certain styles of recreation are inevitably the preserve of a certain class of people in the society—fly-fishing for the professional and business executive, for example, and snowmobiling for the blue-collar factory worker—then to embody one style of recreation in public policy, and to commit our parklands significantly to it, is to yield a valuable and significant public resource to a very limited segment of the population (limited not just by numbers, but by class as well).

The determinist view has been stated most strongly by those whose interest is in humanizing work. "What are we to expect?" the psychiatrist Erich Fromm asks, "If a man works without genuine relatedness to what he is doing . . . how can he make use of his leisure time in an active and meaningful way? He always remains the passive and alienated consumer."[1] Sometimes the point has been put even more strongly: A certain kind of leisure activity is not only to be expected from the alienated worker, but is psychologically necessary for him.

Mass culture reinforces those emotional attitudes that seem inseparable from existence in modern society . . . passivity and boredom. . . . What is supposed to deflect us from the reduction of our personalities actually reinforces it. . . . So, as the audience feels that it must continue to live as it does, it has little desire to see its passivity and deep-seated though hardly conscious boredom upset; it wants to be titillated and amused, but not disturbed. . . .[2]

These observations are a warning to recreational idealists, implying that no effort to encourage more challenging and "disturbing" leisure activity can hope to succeed unless and until the workplace is reformed. The idea is that we observe in present recreational choices a reflection of profound needs that no mere change of attitude or public policy can affect: that those who already have power in the society (like successful professionals) are attracted to recreation that demonstrates to them that they are above needing power; while those who are powerless need nothing so much as to demonstrate (however pitifully) that they are capable of dominion. Thus the distinguished New York lawyer and fly-fisherman lies by the side of a stream contemplating the bubbles, while the factory worker roars across the California desert on a motorcycle.

Though all stereotypes about recreational use are exaggerated, there is some indisputable data. Studies demonstrate very strong correlations between wilderness use and both occupation and education. Blue-collar workers account for only 5 percent of all wilderness visits. One study revealed that two-thirds of wilderness users were college graduates and one-fourth of them had done graduate work.[3]

There is a real irony here. To nineteenth-century thinkers like Olmsted, it was a question of willing our aspirations into existence; and therefore the denial to the ordinary citizen of opportunities for contemplative recreation reflected a decision by those in power to write him off as a hopeless drudge. The modern psychological observers, and the statistics, suggest not only that he writes himself off—but indeed that he cannot help but do so.

While there is wide agreement that the recreation of "the passive and alienated consumer" is to be deplored (recognizing that not all recreation other than the contemplative should be so characterized), it is by no means obvious how one breaks out of the observed work-recreation circle. If alienating work is an important constraint on recreational choice, that only adds one more reason to desire that the workplace be reformed. Does the difficulty in reforming work, however, suggest that it is fruitless to encourage contemplative recreation until there is a social revolution in the factory and the office?

The dilemma cannot be resolved by data, however carefully gathered. The public has to decide how much overt tension it is willing to generate. Edward Abbey, blunt as usual, put it this way in his book *Desert Solitaire:*

> They will complain of physical hardship, these sons of the pioneers. [But] once they rediscover the pleasures of actually operating their own limbs and senses in a varied, spontaneous, voluntary style, they will complain instead of crawling back into a car; they may even object to returning to desk and office and that dry-wall box on Mossy Brook Circle. The fires of revolt may be kindled—which means hope for us all.[4]

A rather less hopeful, but equally provocative, view is presented by Paul Shepard in *The Tender Carnivore and the Sacred Game.* Shepard is describing his childhood in the Missouri Ozarks, and his admiration of the country boys whose hunting was filled with an exuberant "independent, alert confidence" of the sort that Thoreau and Faulkner celebrated as wildness not yet tamed out of men. Yet on returning to see these children as grown men, he finds, sadly, much of the vitality he had known drained out of them:

> The childhood training of hunters is not so much practical training as the opening of spiritual doors by leisured and generous people. . . . Years later, after working in the local factories or on the farm they became dull-eyed and defeated. After they could afford guns they could still get excited about hunting but in a sad way, turned in on themselves, puzzled and querulous.
>
> This is not the defeat of innocence and enthusiasm by age and knowledge. It is because drudgery and toil has blunted them and, worse, their life style had failed them. Hunting had put a premium on physical good

health, on sensitivity to environment and to the nuances and clues in a delicate and beautiful world, on independence, confidence, persistence, generosity, and had given them a powerful sense of the non-human creation. In their adult lives only one of these—persistence—was rewarded; the others were destroyed.[5]

Perspectives such as these reveal strong parallels between the thinking of the modern preservationist and the views that Olmsted expressed more than a century ago. Olmsted saw the average citizen as a victim of aristocratic condescension, and the contemporary park symbolist-preservationist sees him as a victim of industrial alienation. Of course there is a condescension of its own kind in all this, though it must have been perceived quite differently a century ago. The nineteenth-century citizen was told he was being helped to throw off the shackles imposed by a contemptuous upper class. The contemporary citizen—far more committed to a belief in his own autonomy—sees himself characterized by preservationist rhetoric as the prisoner of his own ignorance. Certainly the average park visitor today does not think of himself either as a manipulated puppet or as an externally determined victim. And he does not take kindly to suggestions that his choice of leisure time activity is unworthy. The inability of the preservationist to win a sympathetic majority for his pleas rests on an unwillingness to come to terms with the full implications of his views.

Though the preservationist sometimes appears as yet another critic of mass culture, speaking the language of alienation, he shies away from the more general—and seemingly radical—politics that posture implies. Unlike most mass-culture critics, he seems quite uninterested in social reform. Indeed, most of the time he doesn't appear much interested in people at all. His vocabulary is principally directed to the land and to physical resources, and when he objects to off-road vehicle use or to plans for an urban-style resort in the mountains, his complaint is routinely phrased in terms of adverse impacts on soil, water resources, or wildlife. These are certainly authentic concerns, but they are often viewed as a disingenuous, politically neutral way of objecting to the kind of recreation other people prefer.

As noted earlier, the presence of motorboats in the Grand Canyon is not really an ecological issue, though it was regularly put in those terms. Nor is ecological disruption the sole—or even the principal—reason there has been so much objection to snowmobiles or ORVs. While one element of preservationist advocacy is scientific and truly based on principles of land management, another—and it is very clear in national park controversies—is dominated by value judgments, by attaching symbolic importance to the way *people* relate to nature. When preservationists are condemned for being more interested in trees than in people, there is an edge to the criticism even sharper than it seems. For the impression often given by preservationist rhetoric is that *some* people are less important than trees— the people who enjoy snowmobiles and auto touring and other types of so-called urbanizing recreation.

The criticism is misdirected, but it is an understandable response to the naive and uncandid way in which preservationists often state their position. A more plainspoken statement would be this: The preservationist is an elitist, at least in one sense. He seeks to persuade the majority to be distrustful of their own instincts and inclinations, which he believes are reinforced by alienating work and the dictates of mass culture. To the social reformer his message is that he can help generate incentives that will lead toward reform of the workplace. To those who say "let's look at demand," he says that people need to pay attention to what they ought to want as well as to what they now want.[6] To those who ask how anyone else can purport to know what another citizen should want, he responds that complacent acceptance of things as they are is not the hallmark of a democratic society.

The preservationist's call for a willingness to be skeptical about our own inclinations raises a problem of self-determination that is particularly disquieting in a society deeply committed to ideas of democracy and equality. The concern has been particularly agitating in the context of "elitist-popular" battles over the national parks, but it is actually nothing more than a familiar issue of self-paternalism.[7] A common example is provided by the vacationer who annually brings along a serious book he has long intended to read, only to slip into reading popular mysteries. Similar rituals are familiar in registrations for music lessons or adult education courses, buoyed by the hope that the investment in tuition will be a discipline not to give up, or in the acquisition of sports paraphernalia that collect dust in the attic. Aspiration and conventional behavior are in a continual battle. We are willing to impose coercion on ourselves to some degree (as in paying for lessons that we know we may never pursue) precisely because we recognize that left wholly to pursuit of our routine preferences we are not likely to do and be all that we want. A mixture of autonomy and self-imposed discipline is something we know very well.

Individual behavior patterns have counterparts in public action. Public television is perhaps the most obvious example. We have been willing to coerce (that is, to tax) ourselves to some degree to be induced to view it, even though we know we will probably resist the temptation most of the time. If public broadcasting gave us only what we already knew would be popular, it would simply add one additional outlet to the functions served by the existing plenitude of commercial stations. If, conversely, it was giving us something we knew we didn't want, it would be plainly unworthy of our support. Moreover, public broadcasting cannot be explained simply as a service to the wide diversity of public preferences, for we would never think of offering ordinary public services (like the subway) to as few people as those who constitute the audience for most public television. The most plausible explanation is that we are institutionalizing temptation to pursue some things we have been persuaded we ought to want. At the same time, the pressures we simultaneously generate to make institutions like public broadcasting or the public art museum less highbrow and more popular, demonstrate that there is a tension between self-paternalism and unbridled autonomy that is never fully resolved.[8]

To yield autonomy in this fashion does not undermine commitment to a democratic political philosophy.[9] For if one pursued a philosophy that entirely rejected a willingness to defer to others in giving content to our general aspirations, a very heavy price would have to be paid for that decision. Each individual would have to stand ready to specify his desires exactly. Consider the contrast between the patient who comes to a doctor and orders removal of his appendix, and the patient who asks the physician to help him become healthy again. In the first instance the individual maintains greater control over his own destiny, but at the risk of having to identify for himself exactly what he wants. If one knows what he wants only in aspirational terms (he wants to feel better), then to pursue that aspiration he must give up some of his autonomy. He must let someone else decide in the particular what is good for him, though only in response to something he in general has decided he wants. The problem is even more complicated when aspirations include a desire for opportunities to modify the sorts of things one wants. If avoiding fixity in desires is itself an aspiration, the individual's ability to specify his wants is especially limited. To say "I would like to be a more independent or cultured person," for example, requires even more deference to others than to say "I want to be a healthy person."

In this respect, the traditional question, am I getting what I want, or what someone else thinks I ought to want, may be seen as excessively simplistic. We get both, and the degree to which one or the other dominates depends upon a willingness to accept the possibility that others may know what is good for us better than we do ourselves, on our ability to prescribe for ourselves, and on our willingness to give up some autonomy in pursuit of those needs.

The problem is created not only by lack of knowledge, but by a willingness to confine life within the limits of one's own experience and knowledge. In the simple medical example, the difficulty is merely that the patient lacks certain information or experience. Even there, if he were willing to risk his life by the limits of his own knowledge, he would never have to put himself in the hands of a doctor.

Similarly, in the public context, if the individual were willing to eschew all self-imposed coercion, he could retain a very high degree of autonomy. He could say that libraries should stock only the books he was familiar with, even specifying the titles, or he could establish a university and direct the teachers to teach only what students already knew they wanted to learn.

It is precisely because we have not taken so constricted a view of autonomy that we establish institutions like public broadcasting, or public libraries and universities, and give up *some* autonomy to librarians and professors. The long-accepted presence of such institutions is evidence of our willingness to adopt a political philosophy that yields to others some power to decide what is good for us.

To those who ask the preservationist why he thinks *he* should be the recipient of such deference, rather than any other individual who seeks to lead, he responds that he rests his case on the evidence presented by Olmsted and Thoreau, Cotton and Ortega, Faulkner, Hemingway, Leopold, and the myriad others for whose view of man's relationship to nature he claims to speak. To the extent they are persuasive in stating a general philosophy, he asks the public to accept him as a spokesman before the Congress and the administrative officials who will give these views official status. He recognizes that he has no formal standing. He is at most a member of a loose coalition of people called a movement, and he does not have institutions, certificates, or even an accepted professional or scholarly literature behind him. No doubt this is why there has been a special uneasiness about the aspirations of the preservationist for leadership. But, for better or worse, the preservationist is the only spokesman we have for the tradition of man-in-nature.

Just as the professor of history, or the museum director, speaks for his or her profession, the preservationist boldly asks the public to vest similar power in him—to the extent that he remains within the bounds of his tradition.[10] He asks for something akin to the academic freedom we give a teacher in the classroom, which is not at all a freedom to do whatever he wants.[11] A teacher who assigns a controversial book for his students to read, or a librarian who buys such a book for his collection, will rarely defend that judgment in personal terms. It will instead be urged that the book has attained the status of a classic, that it has been widely reviewed in serious critical literature, or that it is routinely used in college or high school classes around the country. Our response to such controversies is powerfully shaped by precisely such evidence. We may, to be sure, reject every such justification and insist upon the implementation of popular judgment. But if we are willing to give any authority to teachers, as we routinely are, what we give is a freedom to operate within a professional tradition, recognizing that the bounds of that tradition are themselves a significant protection against purely personal or capricious judgments.

As the next chapter will make clear, the preservationist does not have to ask the public to eschew opportunities for conventional recreation except to a quite limited extent. His agenda accommodates to a substantial continuation of ordinary tourism as the routine recreation of most people most of the time. Indeed, more than anything else, he seeks a policy that encourages contemplative recreation as one publicly provided choice, separates it from ordinary leisure time activity, and requires a conscious decision either to accept or reject it.

In these limited, but by no means insignificant, ways the preservationist asks that the public let him lead it. In resting his case on the "evidence" presented by the nature writers, he believes he has a persuasive basis for the deference he asks, and though the tradition is testimonial, rather than scientific, he can at least add to it a range of other documentation that supports the coherence of the philosophy for which he claims to speak.

In a book entitled *Beyond Boredom and Anxiety,* Mahalyi Csikszentmihalyi made a study of activities that generally require much energy but yield no conventional rewards.[12] Csikszentmihalyi's interest was in making the workplace more attractive to employees, and the thesis he set out to test was whether external rewards—such as money and social status—are the only determinants of the incentive to work and of the satisfactions work produces. Pursuing the hypothesis that work could be made more satisfying without the enlargement of external rewards, he set out to discover what induces people to work hard at things they set out to do. He sought out mountain climbers, chess players, dancers, basketball players, composers, and surgeons, and he found some remarkable similarities of response among the participants in these seemingly disparate activities.

Csikszentmihalyi coined the word *autotelic* to describe the common elements he found among the activities studied, joining the Greek words for self, and for goal or purpose. He found that people who feel they are engaged in an enterprise where the goals are self-justifying, in the sense that ultimately the participants set out to satisfy themselves, are able to experience extraordinary levels of satisfaction. The activities he examined are not necessarily autotelic, nor are they necessarily abstracted from conventional rewards. Surgery, for example, has both an external measure of success and an external reward structure. The same is true of certain games, like basketball. The peculiar interest of Csikszentmihalyi's study was the finding that what makes these activities especially satisfying is their capacity to reward the participant according to his own internalized standard. The surgeon self-evaluates a complex operation without regard to the fact that the patient survives, and that he is paid, and he alone knows when he has performed brilliantly, rather than "merely" successfully. In exactly the same way, the serious climber or basketball player has more at stake—in his own mind—than simply attaining the summit or being on the winning side.

The conclusion the author draws from his study is that there are certain kinds of activity that give participants a sense of discovery, exploration, and problem solving, a feeling of novelty and challenge, of opportunity to explore and expand the limits of their ability, that open the way to feelings of profound satisfaction.

Csikszentmihalyi identifies three main elements that underlie the common responses of his subjects. First, a feeling by the actor that he has willingly undertaken the enterprise rather than being induced into it by external incentives or constraints. This, the author suggests, invites a sense of freedom that may be an essential feature of a deeply satisfying activity.

There is also a feeling of being in control in a special sense that makes taking considerable risks acceptable and even comfortable.

Csikszentmihalyi was struck by the distinction climbers made between their voluntary assumption of the risk on the mountain and the risks of driving an automobile. On the highway, they reported, one is vulnerable to the mistakes and recklessness of others; on the mountain one is capable of getting in control of his own destiny. The risks he takes are measured by his own competence and discipline. He is not a passive object of fate determined by others.

Finally, each of the activities has a level of complexity that calls for total engagement, analogous to what was described previously as intensity. The author calls this quality an "infinite ceiling" which heightens concentration and calls for a depth of engagement and a power of perception that is lacking in ordinary activities.

The common element found in all autotelic activities was a range of physical or symbolic opportunities for action that represent important challenges to the individual. Satisfying experiences, the author finds, do not fundamentally involve merely "a passive adaptation to social demands, a normative adjustment to the status quo,"[13] but opportunity for the internalization of satisfaction, based on personal knowledge, individual style and expression, autonomy, and a setting rich enough in its complexity to elicit distinctive personal responses. These conclusions echo strongly the private reflections considered in previous chapters. They suggest the distinction for which Olmsted was reaching in his effort to explain the difference between a visit to old Niagara and the experience of prepared entertainment and mass recreation enterprises that had subsequently overtaken it. The emphasis on personal style and challenge resembles strongly the technique/technology distinction that is central to the recreational literature. And certainly there is a parallel to the contrast Thoreau etched between life at Walden and life in the constrained and constraining atmosphere of Concord.

Csikszentmihalyi's findings and the literature of reflective recreation are affirmed by a major current of psychological literature. Carl B. Rogers, one of the preeminent figures of modern psychology, developed the theory that our most deeply felt (and often deeply buried) need is—in his terms—"to become ourselves."[14] Rogers's observations were drawn from some thirty years of experience as a therapist in which he sought to elicit from his patients an expression of the patient's profoundest feelings about himself—what he is, and what he wants to be.

Rogers's conclusion is that our daily lives are generally bounded by expectations others set for us. Have we behaved as a good parent should? Are we satisfying our employers? Are we dressing, or acting, fashionably? While his patients were eminently responsive to conventional fashions and expectations, Rogers found, they were also often deeply unsatisfied precisely because their behavior failed to answer the question, "who am *I*, and how can I get in touch with this real self underlying all my surface behavior?" He found that his patients began to change and to experience profound satisfaction only as they turned to activity that was founded on their own personalities and their inner resources, rather than on standards set for them externally. People who are able to make this break with convention and to carry it through at the deepest level of their own behavior (people whom the psychologist A. H. Maslow calls "self-actualizing"[15]) come to feel a stronger sense of self-esteem and self-confidence. They have come to terms with their own strengths and weaknesses, and they are less dependent on others for their satisfactions or their sense of self-worth.

At least one familiar strain in the recreational literature, and in Csikszentmihalyi's study, is also found in modern philosophical writing. Professor John Rawls, in his book *A Theory of Justice,*[16] observes that

> human beings enjoy the exercise of their realized capacities (their innate or trained abilities), and this enjoyment increases the more the capacity is realized, or the greater its complexity. . . . Presumably, complex activities are more enjoyable because they satisfy the desire for variety and novelty of experience, and leave room for feats of ingenuity and invention. . . . Simpler activities exclude the possibility of personal style and personal expression which complex activities permit or even require, for how could everyone do them in the same way?[17]

Certainly it would be erroneous to suggest that Rawls and

Rogers, Olmsted, Thoreau, and Csikszentmihalyi are all saying precisely the same things; that all their observations are proven theorems of human behavior; or that taken together they can serve as the basis for confident public decision making. In fact, the existing literature leaves numerous questions quite unsettled. Both Rawls and Csikszentmihalyi, for example, use chess as an example of a complex or autotelic activity; yet chess is the perfect example of the game of conquest and dominance, and in that respect differs sharply from the activities most central to the recreational literature. Indeed, the literature on chess masters is a veritable playground of psychological aberation, with obsessions of dominance and conquest one of its focal points. The great chess players seem to be anything but "self-actualizing personalities."[18]

Perhaps analyses of recreation so far have insufficiently distinguished between those activities that turn on conquest, with inescapable winners and losers, and those that have the capacity to transcend mastery.[19] While there are many important similarities—such as complexity, challenge, independence, and skill development—there seem also to be important differences, not unlike the difference noted earlier between motorcycling and fly-fishing.

Plainly our knowledge about differing recreational activities, the extent to which they are good for us, and whether we ought to want to give them a distinctive place in public policy, is anything but an exact science. And plainly the preservationist has less claim as an authority than does the scientist, the museum director, or the university professor. They put something before us and say, "this is what you ought to want to know," as the preservationist would like to shape a park and say "this is what you ought—if not exclusively, at least importantly—to want of your leisure." But that is the claim he makes, and whether he succeeds or fails in persuading others that he should be followed depends on the strength of the evidence he has. Right or wrong, persuasive or not, his claim is that he knows something about what other people *ought* to want and how they can go about getting it, and he should not back away from, or conceal, that claim.

5

The Compromise Called For

The great bulk of recreational activity will always be of the conventional kind—a day on the beach, a picnic in the park, or a weekend at a resort hotel. The idea is not that reflective recreation should consume all our leisure time, but rather that we should develop a taste for it, and that stimulating the appetite should be a primary function of national parks.

The principal change such a policy would engender is this: Rather than seeking mainly to serve the wide variety of recreational preferences visitors bring with them, park managers would encourage all visitors—whatever their past experiences or skills—to try more challenging and demanding recreation. While the Park Service may believe it is doing this effectively now, the actual pattern of park visitations suggests a quite different conclusion.

Under present practice, a great variety of different areas are available to the visitor, ranging from such easily accessible and rather urbanized places as Yosemite Valley and the South Rim of Grand Canyon to remote backcountry wilderness. In practical effect, however, visitors are channeled to those places that reflect their present preferences and are effectively insulated from settings unfamiliar to them. A first-time visitor to Yosemite National Park, for example, is overwhelmingly likely to find himself in Yosemite Valley, a place full of roads, hotels, stores, and crowds. His opportunity for an experience of contemplative recreation is quite limited, not simply because he has chosen that place, but because that is the face the park almost inevitably presents to someone in his position.

A policy committed to promoting reflective recreation as one of its major goals would focus on the individual who might be willing to try such an experience, but who is neither experienced nor ready to make a substantial commitment. This is the affirmative side of the preservationist's concern about how other people use the parks. The casual or inexperienced visitor may have been attracted by articles or films illustrating the splendors of the forested wilderness or of a wild river adventure. He may be ready to take the plunge if a practicable opportunity is presented.

Such inclinations could be encouraged. Yet the wilderness backcountry is too rugged for him and the popular gathering places too urbanized. To make wilderness areas more accessible, by installing roads there, would put the visitor in the wilderness without exposing him to it, and would also intrude upon others' opportunities to experience challenging wild areas. Places like Yosemite Valley, easily accessible and yet splendid in its scenery and resources, would be an ideal place for such a beginner if it were much less developed and less crowded. The paradox is that an effective policy will not be advanced simply by establishing more wilderness areas, for no matter how much we enlarge the backcountry, and no matter how small the areas devoted to city-type development and motorized nature-loop drives, those latter places will remain the principal magnets for most park visitors.

A related difficulty arises when management commingles the service function with the task of offering novel and challenging experiences. A now-shelved plan to build a motorized tramway to the top of Guadalupe Peak in Guadalupe Mountain National Park in Texas illustrates the problem.[1] The park is situated in a lonely area between El Paso and Carlsbad Caverns and contains the highest elevation in Texas. Jutting out of the surrounding barren country, Guadalupe Peak provides a moderate walk of a few hours to a fine prospect. But the park has virtually no developed facilities and is little visited. A few years ago, the National Park Service published a plan noting that many people want to experience the wilderness, but find it difficult or too time consuming.

The plan therefore proposed the construction of a tramway to carry visitors to the top of the mountain so they could look down into a wilderness area. This, the plan noted, "will be truly a wilderness threshold experience."[2] Peering at a wilderness from a tramway station, however, is *not* a wilderness experience; the sense of wilderness is not achieved by standing at its threshold, but by engaging it from within. Not everyone will seize the chance to experience wilderness, even in the modest dose that Guadalupe Park presents. The opportunity can and should be offered as a choice, to be accepted or rejected; but it should not be falsified or domesticated.

If we want authentic choices, we will have to make some compromises, for we can't have places like Yosemite Valley both as an accessible place for a distinctive recreational experience and as a place to serve conventional tourist demands. One possible compromise is to try fully to serve the quantitative demand for conventional recreation and to provide opportunities for all the different kinds of *activities* the public wants, but not to assure those opportunities in locations that have a special value for reflective recreation. To some extent, of course, we already follow such a policy in the national parks. In general, resorts, and activities like tennis, golf, and pool swimming are excluded from the parks. So long as there is a reasonable opportunity somewhere to participate in all the various activities we want, and with a considerable degree of amenity and convenience, we can reserve critical areas in the parks from conventional tourism without destroying the chance for a conscious choice by the tourist. There would be changes, of course—a reduction in urbanizing influences, with less auto traffic, less densely developed accommodations, a sharp deemphasis on standardized concessioner activities for the novice or bored visitor, and a removal of the resort-type atmosphere that has grown up around some of the park hotels.

To be sure this proposal *is* a compromise, and it does not meet the desires of the visitor who wants to play tennis or go to a nightclub in the shadows of Yosemite's magnificent scenery, who likes the crowds, or who wants to ride a motorcycle at high speed through a desert park. The value of such an arrangement is that it forces us to separate in our minds contemplative from conventional service recreation. Such unbundling of differing wants clarifies the extent to which we are willing to subject ourselves to self-coercion. It requires us to ask ourselves whether we are going to a park because it is a special, challenging, and unfamiliar place without ordinary comforts and services. It puts the symbolism of the parks before us, not as an abstract image, but as a decision with consequences.

One practical difficulty with such an approach grows out of the spectacular increase in recreational use that we have been experiencing in recent years.[3] The demand for service recreation is so great and is growing so fast that it might seem impossible to provide the quantity demanded without imposing even further than we have already on national parks. The problem is not as intractable as it appears to be. Our ability to accommodate elsewhere demands that now press upon national park areas depends significantly on the importance assigned to reserving park areas. An incident in the history of the Olympic National Park (though dealing with commercial demand rather than recreation) provides a revealing example of the relationship between the value we assign to parks and our ability to meet the range of demands that are made upon them.

Olympic National Park in Washington was established largely because of its magnificent stands of Sitka spruce. The Olympic peninsula is also an important area for commercial timber harvesting. Timber companies had long been eager to log the land in the park, and there had been continual dispute over park boundaries. These disputes were relatively quiescent when World War II broke out.[4]

It happened that Sitka spruce wood was peculiarly valuable for military airplane production and was in short supply. Pressures began to mount to permit logging (which is generally prohibited in national parks) in Olympic Park during the war and for the war effort. The War Production Board, whose job it was to assure adequate supplies of material, recommended that timber harvesting be permitted in the park. This situation presented an agonizing dilemma for Park Service officials. They were extremely reluctant to see the park's most distinctive resource impaired. At the same time, both as a matter of conviction and prudence, the official position of the Park Service was one of determined support for the war effort.

The problem was made especially complex by two additional facts. One was a suspicion that the park's lumber might not be needed, and that, to a significant extent, the war effort was being used as a wedge by the local timber industry to get under cover of patriotic need what it had failed to get in peacetime. In addition, the Park Service had bitter memories of World War I, when considerable industrial intrusions had been made on the national parks, particularly for grazing. The Park Service had the good fortune to be supported by a strong secretary of the interior, Harold Ickes. With Ickes's backing Park Service Director Newton Drury issued a public statement saying,

> The virgin forests in the national parks should not be cut unless the trees are absolutely essential to the prosecution of the war, with no alternative, and only as a last resort. Critical necessity rather than convenience should be the governing reason for such sacrifice of an important part of our federal estate.

The policy adopted did not end with a statement of principle, however. Indeed, that was just the beginning. Secretary Ickes himself corresponded with the head of the War Production Board to get a sense of the problem and to make clear the Interior Department's reluctance. It turned out that there was Sitka spruce in both Alaska and Canada, but it had been classified as unavailable. The Alaska timber was remote, and it was doubted that a sufficient amount could be made available in a short time. Canada had put an embargo on the export to the United States of high grade logs, including Sitka spruce, and Ickes suspected that they were holding back production.

With its strong commitment to save the Olympic timber if possible, the Interior Department not only articulated the burden of showing "critical necessity" but set out to find the facts for itself and to suggest alternatives. Park Service employees were dispatched to aircraft manufacturing plants to get a first-hand view of the problem, and to the Forest Service Products Laboratory to investigate alternatives to the use of wood. A proposal was developed for a program of special assistance to private companies logging nonparklands that were difficult to reach, including help in obtaining draft deferments for additional employees. With some inquiry and diplomacy, aircraft log production was increased in British Columbia, relieving the pressure on American forests. Ickes wrote to the War Production Board suggesting eight specific measures to relieve the Sitka spruce shortage without incursions on Olympic National Park.

By November of 1943, at a congressional hearing convened to consider the problem, the War Production Board testified that it no longer believed logging of the park was necessary. A change in aircraft lumber requirements had occurred, the Board stated, following discussions between them and the Interior Department. A decision had been made to construct certain planes of materials other than wood, and an increase in the supply of aluminum available for aircraft production had helped the situation. The War Production Board withdrew its request for park timber and by the end of 1943 the pressures on Olympic had virtually ceased.

The Olympic Park case reveals that claimed conflicts are often less intractable than they appear at first view; that by forcing alternatives explicitly into the open, and by pursuing the facts behind the claims, we can often resolve concrete cases without having to weigh competing values in the abstract. The tension between service of conventional recreation and the preservation of national parks will never wholly disappear, but the problem is not aided by posing questions such as How many acres of wilderness are enough? Like the question of how many books a library should have, or how many Brahms symphonies are sufficient, these are empty canards. If the public accedes to the preservationist position, the task will be to hold on to as much national parkland as other irresistable public demands will tolerate. In dealing with conflict, one must always have a starting point. If the goal is to *encourage* contemplative recreation in the parks, the way to do it is diligently to look for ways to meet other recreational demands more effectively at existing sites, and to scrutinize more carefully claims of need and demand. The strategy is to increase the burden of proof that there is no alternative except the use of parklands: that is the lesson of Olympic Park.

Beyond seeking alternatives, there is a serious question whether the pressures now being felt on public recreation lands are being unduly inflated. Ski resort proposals are among the most frequent and controversial sources of demand for intensive, urbanizing development of the public lands. While they arise much more frequently in the national forests than in the parks, the problem of assessing and responding to demand is aptly illustrated by the ski resort problem. One of the most hotly contested public recreation controversies of recent years arose out of a plan by Walt Disney Enterprises to build an alpine ski village in California's Mineral King Valley, on national forest land just at the southern tip of Sequoia National Park, northeast of Bakersfield.[5] While not a pristine wilderness, Mineral King basin is a beautiful and secluded mountain valley high in the central Sierra range. The valley floor, at an elevation of seventy-eight hundred feet, is dominated by striking peaks rising to more than twelve thousand feet above dramatic slopes. Though there was active mining in the valley at one time, no commercial activity has been carried on recently, and the area has largely reverted to a primitive condition. Good weather, ample snowfall, and fine scenery make it a superlative site for both summer and winter recreation.

The Mineral King proposal was bitterly opposed, and ultimately defeated, by citizen opposition, led by the Sierra Club.[6] The controversy is, on first consideration, extremely perplexing. Why should even the most ardent defender of public lands have objected to the use of the valley for downhill skiing?

Opposition to the Disney plan seemed to be based on some version of the extreme positions that high-quality natural ecosystems should never be degraded, even when they are particularly well suited for skiing; that wilderness hiking or cross-country skiing should be preferred to alpine skiing; or that the recreational preferences of some public constituencies should simply be given priority over others. The controversy was made particularly baffling by the fact that the principal opponent of the Disney plan, the Sierra Club, was the very organization that had first suggested the site. The problem arose this way.

Substantial growth in Southern California's population during the mid-1940s threatened to make inadequate the available facilities for downhill skiing within easy reach of Los Angeles. Ski enthusiasts urged the development of the nearby San Gorgonio area, which was within a national forest. But because the area had been classified primitive—a term the Forest Service used to reserve land for wilderness-type recreation—conservation groups, including the Sierra Club, rejected the San Gorgonio site.

Both the Forest Service, which traditionally played an entrepreneurial role in facilitating recreation development on national forest land, and the Sierra Club, which at that time routinely accompanied its opposition to any given site with efforts to find a more appropriate location, suggested the more remote Mineral King Valley. The Forest Service then sought bids from private ski developers, as was its practice, but no responses were forthcoming, principally because access to Mineral King required the improvement of a road into the valley. The cost of the new road was so great that it would have made commercial development economically impracticable.

For some years the idea of developing Mineral King lay dormant, but by the mid-1960s Disney Enterprises, through adroit political effort, had obtained agreement by the state of California, with aid from a federal government agency, to finance the access road. Disney then asked the Forest Service to renew its request for bids to develop the valley, which it did. Disney put forward a development proposal which was accepted. At this point, however, the Sierra Club decided to oppose construction of a ski resort at Mineral King. The issue became a celebrated national controversy; the Sierra Club sued to prevent the development, taking its case to the United States Supreme Court.[7]

Having originally suggested the Mineral King site, the Sierra Club could hardly avoid some embarrassment at the subsequent vigor with which it fought the Disney proposal. There are many explanations for this shift of position, all of them no doubt accurate as far as they go. It is true, for example, that environmental consciousness was sharply on the rise during the time the new Mineral King development was urged; that a new urgency was given to the preservation of remaining wilderness; that the Sierra Club had itself changed character, becoming more preservationist in its outlook; and that the Disney plan was considerably more ambitious than anyone (including the Forest Service) had anticipated.

The accuracy of these considerations does not, however, adequately explain the intensity of opposition, and the symbolic meaning, that the Mineral King battle took on. For example, the developer's promise that modification of the natural resources would be minimized to the fullest extent consistent with a ski facility did not reduce the opposition,[8] for it soon became clear that the opponents were not interested in a well-developed ski resort, but wanted no ski resort at all. Even a former Sierra Club president was unable to support, or find a justification, for such opposition. Breaking ranks with his colleagues, he said: ". . . [T]he American way of life . . . is pluralistic . . . [I] ha[ve] no use for the argument that there was something superior in the wilderness use and that the skiers should be considered a second-class use."[9]

His statement is perfectly appropriate, but like many participants on both sides of the battle, he failed to see through to the real issue. The controversy over the proposed Disney ski resort was not a disagreement about skiing and its place on the public lands, but about resorts, and *their* place on the public lands. The essential facts were not the physical impacts on the land, and the possible minimization of those impacts, but rather that Disney was proposing to invest thirty-five million dollars in a facility that would accommodate some eight thousand people a day in a valley of about three hundred acres; that within this area there were to be thirteen restaurants, seating 2,350, including a 150-seat coffee shop at the top of an eleven-thousand-foot mountain to which a high-capacity gondola was to lift visitors; that there was to be parking for 3,600 cars, twenty-two ski lifts, and, in addition, a full complement of swimming pools, a horse corral, a golf course, tennis courts, and a considerable number of shops.

The central issue at Mineral King was not simply service to skiers who needed such a site, but the development of a magnet facility calculated to draw to it as many people as possible. The Disney proposal was really a plan of the sort we encourage for central city malls, where the idea is to provide so wide a variety of attractive activities that people are strongly induced to come to the place, even though they may previously have shunned it. In the mall situation, our goal is to attract crowds to the city, to make the place a lure for activity because we want people in that location and are quite indifferent to what they do once they get there. But there is no public advantage as such in attracting crowds to places like Mineral King Valley. If we are to draw people to the valley, it should be because the place provides a special opportunity for a distinctive kind of recreation; or it is a needed site, because of its physical characteristics, for activities that cannot reasonably be accommodated elsewhere.

The Mineral King dispute nicely illuminates a difference between an entrepreneurial and a public policy perspective on dealing with questions of recreational demand.[10] Plainly the entrepreneur is doing more than filling a demand for ski opportunities. He is trying to fill the place where he does business. Mineral King presented the Disney version of a policy for public mountain lands, and, unfortunately, that version has insinuated itself into public policy as well.

The dynamics of resort development on the public lands was recently further illuminated by a proposal to develop a facility called Ski Yellowstone in Montana's Gallatin National Forest.[11] There, because the Montana Wilderness Association had filed a competing application to develop a nonprofit cross-country ski facility, and because neighboring ski resort owners, themselves having financial problems, submitted comments on the Ski Yellowstone proposal, the record on the application for a permit to use national forest land provides a rich store of information about such developments.

In commenting on the Ski Yellowstone proposal, the manager of another ski area in the region—Jackson Hole Ski Corporation—explained that his company had been in business for twelve years, and only in the last two of those had it earned a profit. His explanation was that "only in the past one or two years has [our company] reached the point of having enough restaurants, shops, cocktail lounges and other activities to satisfy the destination resort customer for one week."[12] The resort operator needs a critical mass of people to support the facilities that in turn support him. As the owner of the Sun Valley ski resort in Idaho explained: "[t]he village had to have a certain size. You had to have enough people to support five or six good restaurants, and more than one night spot."[13]

The scale and proliferation of ski resorts on the public lands reflects these business demands. The Grand Targhee Resort, not far from Yellowstone, noted that it could not operate profitably without a minimum of six hundred skiers per day;[14] and the Jackson Hole resort said that it was only when it began to fill overnight facilities for two thousand visitors, accommodation for twenty-six hundred skiers a day, six hotels, and twenty small businesses that it began to show a profit.[15] The Forest Service's own study of the ski industry observed that areas like Colorado's Telluride and Big Sky in Montana would be unlikely to get over the problems of low profitability and low utilization until they completed the development of a "full range of resort facilities."[16]

The Ski Yellowstone controversy makes clear that meeting demand, in the simplest sense of the words, is only a modest part of what underlies such applications. The entrepreneur is not merely meeting demand but stimulating it; or, if one prefers, he is tapping latent demand. The concept of latent demand, however, raises a number of questions for a public policy. How do we know whether there is more latent demand for this activity than for some other? How do we know what latent demand, and how much of it, we are tapping? To what extent are we attracting skiers, and to what extent are we attracting those who simply want a winter resort?[17]

From the point of view of public land policy, these are questions of considerable importance. If we adopt a policy of encouraging reflective recreation, a resort setting—where a primary goal is to entertain and occupy the visitor—is likely to conflict with the policy. If we are being asked to meet demands for certain services, then the question is how much the other policy must necessarily yield. If the need is only pressing for ski facilities, then such facilities (but without resort accoutrements) might well be provided without intruding on the capacity of the area to sustain reflective recreation. If there are people who want skiing, and some resort-type service along with it, we ought to ascertain just how much such demand there is—and then decide to what extent we want to provide such a mixture in any given area. A number of quite distinct questions need to be unbundled.

The entrepreneur, however, has an entirely different perspective. The less he unbundles these questions the better, for he wants as many people as possible to be attracted to his resort. Moreover, in one important respect, his goals are in direct conflict with the public policy proposed here. The entrepreneur has a positive goal of attracting a clientele interested in visiting shops, restaurants, bars, and nightclubs.[18] He makes his money from spenders and crowds, not from those who are seeking in a solitary way to find their own style. As one observer noted (talking about a similar problem in regard to snowmobiling):

> The motel owners have tried to operate year-around. . . . [W]e got together to see if we couldn't organize something around snowmobiling. . . . In 1969, we would have one motel, a restaurant and two grocery stores open on a winter weekend—and that was all. Now we have about 20 motels, four service campgrounds. . . . When you talk about economic value, snowmobiling is unsurpassed. It's an expensive sport and spreads a lot of money around. The hikers bring their own food and stay in an Appalachian Mountain hut. Snowmobilers stay in local motels, eat at the restaurant, patronize the gas stations.[19]

These pressures are less evident in the national parks, where resort-type development has in principle been resisted and where concessioners are traditionally more tightly restrained. But the pressures are there nonetheless, and at times they rise dramatically to the surface, a phenomenon that has been intensified since large recreational corporations have begun to acquire the traditional small, local concessions in the parks.[20]

The most celebrated recent incident occurred shortly after the giant entertainment conglomerate, Music Corporation of America (MCA) acquired the concession in Yosemite National Park. Soon afterward, a proposed master plan was issued that incorporated much of the development plan suggested by MCA.[21] MCA's proposal included expansion of overnight facilities in the already crowded Yosemite Valley, replacement of rustic cabins with modern motel units, construction of an aerial tramway from the valley floor to Glacier Point for "viewing, eating and sundry sales," additional concessioner-operated back-country camps, a proposal to stock park streams with nonnative trout, encouragement of winter recreational use and expansion of visitor facilities at Hodgdon Meadows, promotion of convention business in the park, and a number of other developmental proposals.[22] Only after an anguished public outcry, and a series of sharply critical congressional hearings on the MCA efforts (including the filming of a television series that included helicopter ferrying in the park) did the Park Service undertake a thorough reconsideration of its management plans.[23]

This was an extreme case, to be sure, and it arose at a time when the Park Service had its least distinguished and least professional leadership in the directorship, but the problem is a perennial one.[24] It crops up even in such minor matters as acquiescence in park souvenir shops selling baubles with no relation whatever to park resources, defended on the ground that such concessions are necessary to keep the shops profitable.

A policy of unbundling demand could considerably reshape use pressures on the public lands. To allow the development of ski facilities, but discourage accompanying resort facilities, would provide a service to those whose principal interest is skiing. It would permit the continued development of places like Utah's Alta in the Wasatch Mountains, a sparsely developed ski facility which has operated for many years with a modest lodge, a pleasant restaurant, and superlative alpine skiing.[25] It would divert those who are looking principally for a winter resort vacation to private lands or to places on the public lands that have less value for their undisturbed natural features. It would thus minimize the sort of conflicts that arose at Mineral King, while leaving such places available for modest scale downhill skiing developments if the demand for skiing cannot reasonably be met elsewhere.

To adopt such a policy would reduce some of the current conflicts over recreation lands, for a significant share of the present demand for such mixed experiences as resort/skiing developments is surely being generated by entrepreneurial initiatives in search of a way to fill vacation time. Such needs do not necessarily call for ecologically or scenically important public lands. Some years ago, Brian Harry, who was the chief naturalist at Yosemite National Park, remarked: "People used to come for the beauty and serenity. Those who come now don't mind the crowds; in fact they like them. . . . They come for the action."[26] If it's just the "action" that is being sought by many, and there is a plenitude of nonskiing "action" at many ski resorts, the demand can be diverted elsewhere.

The policy suggested here would no doubt produce a decline in the promotion of skiing by big industrial enterprises, such as the giant corporations that tried to develop Big Sky in Montana, and the airlines that have latched onto skiing resorts as a major air vacation destination. Much public land resort development would be exposed as economic pump priming for local communities, designed to generate road, airport, hotel, and restaurant development and accompanying employment. It would also be revealed that a number of ski resort proposals are little more than vacation-home developments in public land settings, now taking the form of the burgeoning condominiums to be seen adjacent to many ski facilities.

The accommodations and compromises proposed here, however, do not exhaust the problem. For if urbanized recreation continues to grow at present rates, a commitment to meet it could well overwhelm any effort at compromise, even with the best efforts to implement the Olympic Park approach and to separate different kinds of preferences. If the policy proposed here is to succeed, it will have also to moderate total demand for the kinds of conventional recreation that are most in conflict with it. The most serious practical problem in meeting current recreational demand is presented by those activities that are highly consumptive of resources: high-powered vehicles that require a great deal of acreage; noisy motors that create conflicts with other uses over a large area; hurried visits to a multitude of places, creating crowded conditions; uses that exhaust large quantities of energy and demand substantial development of facilities. This is not a description of all conventional recreation, of course, but of the types of such recreation that are most at the center of controversy. Such activities are characterized by intensiveness of resource use, or intensiveness of consumption, rather than by the intensiveness of experience that defines the preservationist ideal for use of the parks.

The significance of such intensive-*use* activities is not merely that they impose unprecedented pressures on a limited base of resources, but that the satisfactions they produce are directly correlated to the increasing exercise of power and of consumption. If a motorcycle is good, a more powerful and faster cycle is better. If a resort is good, a bigger resort with even more things to do is an improvement. Unlike some ordinary tourist activities (a picnic or a volleyball game) which are simply different from reflective recreation, power-based recreation is antithetical to it. The fly-fisherman, for example, simplifies his tools in order to reduce power over his experience. The consumer-recreationist does precisely the opposite.

If the preservationist does not succeed in reducing the taste for such activities, he will fundamentally have failed. His goal is to encourage the public increasingly to internalize its capacity to wring satisfaction out of experience—not merely for the brief moments spent in the parks, but in the attitudes carried away from them as well. In this respect recreation policy fundamentally reinforces the symbolic value that parks embody for the preservationist. As symbols of restraint, and human limits, their message is inevitably undermined unless they affect the attitudes we bring to the use of our leisure time.

Indeed, the issue is not simply reducing conflict between opportunities for different kinds of recreation. It is unlikely that we could fill the exploding demand for power-based recreation even if that were our first priority. Recreation that is dependent on ever-increasing growth and impact for its satisfactions is insatiable. The scarcity of resources we encounter in trying to meet such recreational demand is as much a psychological as a physical problem. No matter how much land we have, more will always be demanded because the object is itself more, more of whatever there is. This is what the Spanish philosopher José Ortega y Gasset called "the psychology of the spoiled child"[27] who is insatiable because his object is not some particular thing, but a larger *share.* Increase is itself the object of his desire.[28] This, perhaps, is another way of asserting that the will to power is ultimately self-defeating, and that the preservationists' moralistic stance may be a practical solution as well, even for those who can only see the problem as one of perpetually insufficient physical resources.

The distinction between recreation that draws on intensiveness of experience and that which draws on intensiveness of impact is analogous to that between serious literature and commercial mass entertainment. In the former case there is never a shortage of material; a reader cannot in a lifetime begin to exhaust the available resources precisely because the material's capacity to engage us turns on the intensity of experience it demands. In contrast, commercial entertainment is chronically short of materials. It uses up writers and stories at a furious rate, and it finds itself drawn to material of ever-increasing impact—more violence, more sex, more shocking situations—to maintain the viewer's attention. It is feeding an appetite that, based on external stimulation, grows more the more it is fed. It generates its own scarcity.

Our recently increased appreciation of the physical limits of the world to provide goods and services in boundless measure to everyone—our concern about crowded land, shortages of energy and minerals—give weight to Ortega's observation that so long as we continue to believe in the principle of increase as the measure of satisfaction of our desires, we will never be satisfied and will never avoid scarcity.

The parks themselves, however they are used, will never constitute more than a small fraction of all our recreational resources.

And ideal forms of recreation will never account for more than a tiny fraction of anyone's leisure activity. But the underlying idea—substituting intensiveness of experience for intensiveness of consumption—can radiate out into a much wider area of both private and public recreation and can speak broadly to the problems of scarcity and conflict that we see everywhere. Power-based recreation will continue to present limitless demands until we come to terms with the implications of power as a recreational motif.

6

The Parks as They Ought to Be

Nothing turns upon whether visitors are hiking or riding, young or old, staying in a hotel or sleeping in a tent. The hotly contested question of vehicle use in the parks, for example, is not an issue of transportation, but of pace. Intensity of concentration on the natural scene and attentiveness to detail are simply less likely to occur at forty miles an hour. For *this* reason it is appropriate to discourage motorized travel. Such a policy would not militate against all road building in reserved parklands. We need reasonable access to the various areas of very large parks. And because reserved lands should affirmatively be made enticing to as wide a spectrum of the public as possible, including newcomers who need a taste of the opportunities the land offers, it makes sense to have—as we do in many parks—a highway designed to provide an introduction for those who are deciding whether they want to come back for more.

The purpose of reserving natural areas, however, is not to keep people in their cars, but to lure them out; to encourage a close look at the infinite detail and variety that the natural scene provides; to expose, rather than to insulate, so that the peculiar character of the desert, or the alpine forest, can be distinctively felt; to rid the visitor of his car, as the fisherman rids himself of tools.

The novice, the elderly, and the infirm, as well as the experienced backcountry user, can all be embraced within the same policy. Those who have little vigor or impaired capacity may be limited to a smaller area or to less grueling terrain, but there is an abundance of experience to be had within easy reach in any complex natural ecosystem for those who are willing to trade intensiveness for extensiveness of experience. Indeed, the more immediately nature is met, the less total land that is required.

The concern that has been expressed for the elderly and the infirm in debate over parkland developments must be taken with a measure of skepticism. People who were active when they were young ordinarily continue to be as active as they can when they get older, and those who are reluctant to leave their cars range widely across age groups. Neither the elderly nor the infirm, if they were active at other times, are in the forefront of those advocating intense development of parklands. Rather, those who urge development have put the elderly and the handicapped on their front line. I have myself climbed in Montana with a fifty-seven-year-old totally blind man, who was continuing—to the best of his ability—to pursue the kind of activity he enjoyed before his injury. And I have walked down and back up the Grand Canyon with a husband and wife in their late sixties, who at a slower pace were repeating adventures they had previously savored.

Management committed to contemplative recreation should be just that, whether for the young and hardy or the old and infirm. One does not provide such an opportunity for older people or inexperienced visitors by building a highway to the top of a mountain. Rather we can assure that places that *are* accessible to them are not so deprived of their natural qualities as to put such an experience beyond their reach. If it were necessary to go into the rugged backcountry before finding a relatively undisturbed ecosystem, the lesson would be that we had too ruthlessly developed the more accessible places, not that still more places should be deprived of their complexity.

Converting such ideas into specific management practices is often a subtle task where tone and suggestion are the critical factors. Fixed boardwalks in very fragile and accessible areas, such as the Everglades's Anhinga Trail, may be the only option practically available. Even where some guidance is needed, it can be provided with imagination and subtlety, rather than being reduced to drearily fixed tours, with visitors taken in lockstep from place to place at a predetermined pace. The National Audubon Society's Corkscrew Swamp Sanctuary in Florida shows what can be done. Though the physical nature of the swamp requires the presence of established walkways, visitors guide themselves through the sanctuary. Interpreters are stationed at various significant places within, available to help the unsophisticated visitor see things he might otherwise miss, but they are never intrusive and they do not determine the pace or the quality of the experience. Everglades National Park has a similar arrangement.

A motorized nature-road loop in a park like Great Smoky Mountains, on the other hand, is a dubious facility. The Roaring Fork Motor Nature Trail in the Smokies, set at the edge of the park nearest the tourist town of Gatlinburg, serves as a magnet for precisely those casual visitors who could be invited to penetrate the park on foot. Yet it serves just the opposite function, offering a vista with a fixed beginning and end and implicitly encouraging the visitor to remain in his car, rather than inviting him to see the Smokies at a reduced pace and at close range.

Perhaps the Park Service sees such places as desirable diversions, isolating the casual visitor who simply wants to say he has been in the Smokies without intruding on others or threatening the park's natural resources. If so, it is a step in the wrong direction, for the inexperienced, urbanized visitor is precisely the one who needs the most attention and on whom the most imagination needs to be expended. Magnet facilities are required for them, but a better attraction would be a short access road leading to a variety of trails which should, where possible, be of indefinite length, with opportunities to cut back at various points, in response to the visitor's own inclinations, but without suggesting that the walk has come to a decisive end. Even the least active visitor could be accommodated by such a system. If even this incentive does not work, then, and only then—but at that point decisively—the Park Service should tell the guest that he has come to the wrong place.

The Great Smoky Mountains National Park is an important

testing ground because it is one of the most visited facilities in the system, is close to populous eastern and midwestern population centers, and is subjected to strong pressures for urbanization. In general, it has been admirably managed. The vast bulk of the park has been retained without development and the Park Service has spent considerable effort in closing old roads and resisting demands for new ones.

The Smokies also provide an instructive example for the problem of lodging. There is no hotel in the park accessible by automobile, yet there is a facility for those who are not prepared to camp in tents. The Mount LeConte Lodge, deep in the center of the park, is a simple group of cabins with a dining facility and lounge room/office nearby. No roads reach the lodge; provisions are packed in by horse, and visitors walk in by several different trails, each about seven miles from the nearest parking lot. The walk has moderate rises in elevation, but is easily accessible in less than a day by even the most casual hiker. The lodge is frequented by many families with small children and by a good many people well beyond youthful exuberance. Its accommodations are limited to about forty visitors, and it provides no amusements. There is nothing to do when one gets to Mount LeConte except what the visitor finds for himself. The Lodge is an invitation to find out what the park offers, accommodating the relatively inexperienced visitor without lapsing into the familiarity of the conventional resort. While the Park Service has had its problems with the concessioner at Mount LeConte—and has wanted to close the lodge—that facility embodies a concept of affirmative service to the less experienced visitor that the parks should not reject.

Happily, the lodge has not expanded to accommodate all the visitors it could draw, and in that respect it illustrates an important point: Crowds diminish the opportunity for visitors to set their own pace. It may be said that millions of people want to visit these places, and that no one should be denied the opportunity. True enough. Yet it is impossible to provide unlimited visitation and the essential qualities of an unconventional, nonurban experience simultaneously. Here too a compromise is called for: a willingness to trade quantity for quality of experience.

There is nothing undemocratic or even unusual in such a trade. The notion that commitment to democratic principles compels the assumption of unlimited abundance and a rejection of the possibility of scarcity is one of the familiar misconceptions of our time. We need a willingness to value a certain kind of experience highly enough that we are prepared to have fewer opportunities for access in exchange for a different sort of experience when we do get access. We already ration backcountry permits in the national parks in order to avoid crowds in the wilderness, and we even ration overnight access to such unremote places as the Point Reyes National Seashore near San Francisco, or the Año Nuevo California State Park where the elephant seals visit to breed during their annual migration. The private Huntington Library near Pasadena, California, now limits by reservation the number of visitors who can enter its splendid and tranquil gardens on Sundays, thus maintaining the quality of the visit at the expense of numbers. Some states ration big game hunting permits, for which individuals may wait a number of years in order to enjoy one remarkable hunting experience. Indeed, the experience may be more highly valued because of this. The visitor's sense of anticipation is heightened, and entry to the place made more dramatic by "rationing." In all these devices there is equality in the right of access, but a reduction in the total quantum of access in order to exalt quality over quantity of experience.

The challenge of providing an unconventional experience is greatest in those areas where the growth of the national park system has itself been greatest in recent years—at the island and shoreline parks established as national seashores and lakeshores and at the urban parks created within major cities. These places present a number of special problems. They are not, like the early western parks, vast regions of undisturbed wilderness. New York's Fire Island, for example, encompasses existing vacation communities with homes and stores and a building boom in its midst. Indiana Dunes National Lakeshore is literally in the shadow of the belching Gary steel industry and has to cope with a nuclear power plant being built at its border. The Gateway parks in New York and San Francisco are within bustling cities. The Sleeping Bear Dunes National Lakeshore in Michigan consists of a series of unjoined pieces held together by a long-standing vacation home community. In general these places have had to be carved out around private developments, preventing the isolation that has traditionally allowed parks to create their own ambience.

The risk is that these places will simply become appendages to the communities where they have been created; that homeowners will see the parks as enlarged backyards, or as extensions of city parks, distinguished only by the fact that the federal government finances their management. The newer park areas are already viewed by many area residents as unwelcome intruders, suppressing a hoped-for bonanza in real estate values.

New urban area parks cannot feed on the traditional symbolism of wilderness, but they have a rightful contemporary place in the national parks system, for even with the advantages of modern transportation and affluence a great many people will never likely visit remote jewels like Glacier National Park or the Gates of the Arctic in Alaska, or even Yellowstone and the Everglades. The growth of the national park system is justified by a recognition that the symbolism of parks needs to be brought closer to the public, not that the symbol should be urbanized.

The urban-region park provides an ideal opportunity to show city dwellers that the psychology of the spoiled child is not the only choice open to us; that we can draw satisfaction by accommodating to natural forces as well as by harnessing them. By refraining from driving roads into every corner of every park, as the Congress recently did at Assateague Island National Seashore,[1] even a small acreage can be made capacious. Places become much bigger when we are on foot, and a slower pace enlarges the material on which to expend our leisure.

By having some places where structures are not built along fragile ocean shorelines, we provide ourselves with an object lesson in the economies of accommodating to the forces of nature—and we can see, by contrast, the cost when we close our eyes to these matters. Early planning for Gateway National Recreation Area in New York's Jamaica Bay was built upon a recognition that park visitors live in an irrevocably man-made environment most of the time. Thus it set as an objective to

> forge an effective link between the urban values systems that characterize communities in New York and New Jersey and the natural systems of Gateway . . . [to] dramatize for the public the gains which can accrue if swimming and shellfishing are enlarged . . . and use Jamaica Bay as proof that modern man can work with nature and reclaim what has been impaired.[2]

The Gateway plan also called for energy supplies and new techniques of resource recovery in the management of the park—the latest solar, wind, and waste systems—that would provide "a working exhibit and testing ground of efforts of man to live in harmony with nature."[3] By illustrating the energy efficiency of natural systems, and employing energy efficient innovations in its own management and transportation regimes, the urban parks can show that a national park is neither a place to which one escapes from the reality of the world, nor a place to which one brings its conventions, but a unique facility with important ideas and experiences germane to our everyday lives.[4]

Most important of all, the new, closer-to-home parks provide an opportunity to show the urbanite what it means to be without what Olmsted called distractions. Though Olmsted's Central Park of the 1850s was not a natural wilderness (indeed it was extensively and cunningly landscaped), it was to be a place without amusements and hawkers to amuse the masses. Olmsted's goal at Central Park—lamentably later betrayed in many instances—was to put the New Yorker into a setting designed to stimulate his imagination, and then leave him alone so that his own inclinations and thoughts could take over. With the new facilities of the national park system in New York and San Francisco, Cleveland and Los Angeles, and a number of other places accessible to the great majority of the public, rich and poor, we have a second chance to realize Olmsted's vision.

Setting one's own agenda is one of the most difficult ideas to convert into a set of administrative directions. Yet many of the values articulated earlier, such as freedom from conventional expectations, developing a distinctive personal style, coming to terms with inclinations toward domination and submissiveness, and reaching for the possibilities of boundless involvement, ultimately depend upon setting one's own agenda.

At the simplest level, this means avoiding conditioned responses that get in the way of freshness of experience. Again, the Great Smoky Mountains National Park provides a useful example. At one time areas along the roadways were carefully cut and trimmed, creating a lawnlike appearance. When a new superintendent was appointed, he ordered this practice stopped, which engendered a good deal of complaint from visitors. The roadsides had been so attractive, they said, so neat, and now they had a rough and ungainly appearance. On this small but significant point the superintendent was adamant, however, and for exactly the right reason. Visitors to the park were reacting to a conventional, familiar, and deeply ingrained image of beauty—the trimmed and landscaped lawn. The goal should not be to stimulate that familiar response, but to confront the visitor with the less familiar setting of an unmanaged natural landscape.[5] The mild shock of a scene to which there is no patterned response, and the engendering of an untutored personal response, is precisely what national park management should seek, even in such seemingly small details.

Exactly the same point might be made about wildlife. Those who came to certain parks loved the housepet quality of the bears who came down nightly to feed at garbage pits that were provided for them in full view of park visitors. This was the familiar Hollywood version of the wild animal, a picture which virtually all of us carry around in our heads. We like it and we respond conventionally to it—the cuteness, the docility, the anthropomorphism of the housepet.[6] But the parks can put the visitor in contact with quite another version: the animal in his own habitat, behaving quite without regard to any predetermined notion of how we would like him to behave, sometimes threatening, almost always elusive, at times quite annoying. When we have to react to park animals in this setting, we are on the way to making our own agenda.

These are simple and obvious examples. The agenda issue has its subtle side as well. In one respect it is certainly true that no one requires the visitor to spend his time in any particular way. There is no social director, as at a resort or on a cruise ship, pulling guests by the arm to do this or that. Nor is there an obtrusive program imprinted on the landscape as at a Disneyland, a completely self-enclosed world where the management affirmatively sets the agenda—moving us along cleverly designed paths from ride to ride and restaurant to gift shop, stimulated by upbeat music and bright colors; the clean and happy world of Wild-West saloons and smiling plastic alligators.

For the park visitor who is able, and who knows what he is looking for, there are no such constraints. He can go off and find what he wants in the outer reaches of the place. But for a great many visitors—for a considerable majority, I have concluded from frequent observation—the agenda question is very much less settled. They come to the famous national parks because it is widely believed that these are among the very special places in the world, in the sense that the major cities or the great European cathedrals are places well worth a voyage. There is a sense that something special is to be found here beyond the routine of daily amusement, something with the power to enlarge, to stimulate, to capture the imagination.

So we come. And the experience is initially stunning. Like walking for the first time into Notre Dame or the Sainte Chapelle of Paris, there is a sensory shock in seeing the redwoods, the Grand Tetons, or Mount Rainier that dazzles all but the deadest souls. Yet the initial experience is not long sustained when it is nothing more than amazement at a stupendous visual prospect. I recall a young man working as a waiter at the El Tovar Hotel in Grand Canyon National Park. After six weeks he was getting ready to leave. "How long can you keep looking into that hole?" he asked. "There isn't anything to *do* here."

It is at this point that subtle, but vital, questions in administration arise. There is every reason for a park to have hotel facilities for those who do not wish to camp in tents, though Mount LeConte Lodge is a better model than Yosemite's

Ahwanee, with its conventions, stuffy elegance, and souvenir shop selling pinecones in cellophane bags. Nor is there any reason to abolish campgrounds suitable for those who do not wish to pay, or cannot afford, nightly hotel prices. Neither need we root out high country camps or trail shelters, where these serve an interpretive function or give protection from particularly heavy weather (though such facilities have a tendency to become garbage dumps and gathering spots for spiritous as well as spirited socializing). Supportive services—supply stores, unpretentious restaurants associated with hotels, and gas stations in more remote parks—are also perfectly appropriate. What do not belong in such places are facilities that are attractions in themselves, lures that have nothing to do with facilitating an experience of the natural resources around which the area has been established.

For example, the Jackson Lake Lodge in Grand Teton National Park is a lovely resort hotel, but it disserves the sort of opportunities the park ought to be stimulating. It is an attraction in itself, with its fancy shops, swimming pool, and elegant restaurant. Obviously one need not stay around the hotel using it as the centerpiece of a visit, but to the extent it attracts visitors, it discourages setting one's own agenda within the park's natural resources. Such overdeveloped facilities are unnecessary. We can go to resort hotels elsewhere. If we come to experience the Tetons we should be willing to recognize we are in a distinctive place offering an unfamiliar experience we must search out for ourselves. This is not to suggest that such places do not have their virtues. The new, highly developed balcon resorts in the French Alps, for example, are in many ways highly attractive places at which to be instructed in mountain skiing. As one observer noted, "things are superbly organized. The lifts are close by, the slopes well groomed, ski classes are held on schedule. The balcon resorts are great, efficiently organized machines."[7] No doubt that could be taken as a compliment but it is hardly the ideal metaphor for a national park.

This, of course, is a matter of setting a tone for a place, but creating the appropriate tone is very much at the heart of the matter. The problem is not hypothetical, as even the most casual observer will notice. In the midst of a recent summer season, Jackson Lake Lodge was largely given over to a business convention. Hundreds of people, dressed in resort style, were continually flitting in and around the hotel lobby, going from meetings to shops to restaurants and bars. Doubtless a number of them carried away from their visit a heightened interest in the park's magnificent resources, but certainly they would have done so no less if other amusements and attractions were removed. Undoubtedly some would never have come to the park except for the convention, and the convention might not have gone to the Tetons in the absence of a package of resortlike facilities. But this is just another version of the bundling problem. If a place like the Tetons cannot attract someone based on its own resources, then that visitor may not be ready for an encounter with nature. Like the museum or the university, the park can wait for patronage until the aspirational urge is in the ascendency. No doubt some people will miss an opportunity to which they would have responded if only they could be lured inside by other means, but to achieve this purpose requires a blurring of mission that—for the reasons set out earlier—is best served by keeping different management goals as distinct as is practically possible.

The other major problem of this kind has to do with the concessioner who offers boat and horse and bus rides, and the like. In theory, there is certainly nothing wrong with such activities. One ought, for example, to be able to rent a horse or a boat in a park. Yet frequently the commercial imperatives of a concessioner reshape these services in a way inconsistent with the demands of management for self-defined recreation. One must rely upon impressionistic evidence, and my strong impression is that a great deal of the activities of concessions on the public lands are designed precisely to fill a void for the visitor who has come expecting to be entertained. For that reason they cut against a policy calculated to force the agenda question back upon the visitor.

Certainly there is a banality and predictability about many of these functions that has little to commend it. The drearily routine mule rides at the South Rim of Grand Canyon for which people line up morning after morning; the one-hour, two-hour, four-hour, horseback loops, with a daily breakfast ride or "chuck wagon dinner" thrown in, that are so common a sight; the round-the-lake commercial boat ride that is a standard feature at a number of parks. All these are nothing but amusements, however beautiful the setting, and they seem indistinguishable from the local pony ride. Their capacity to get visitors deep into the park experience seems minimal, they have a mass production quality about them, and they have a considerable capacity to detract attention from the fashioning of a personal agenda. They can be dispensed with.

7

"At the Core of All This Wilderness and Luxury"

Until the early 1960s boating down great untamed western rivers was virtually unknown. Except for a few hardy individuals who took the peril in kayaks and canoes, or in boats they fashioned themselves, the prospect of facing extremely rapid currents, steep falls, and huge waves was too much to permit the development of anything like a mass form of recreation. Wild rivers—the Colorado in Grand Canyon or the Middle Fork of the Salmon in Idaho—presented the challenge of white-water boating carried to its ultimate possibility.

Then a series of things happened to bring about change. Public attention began to be focused much more sharply on wilderness in the years leading up to enactment of the Wilderness Act of 1964.[1] Among the areas that received the most attention were the spectacular canyons through which these wild rivers ran. When proposals were made for the development of hydropower dams in the Grand Canyon area itself, the Sierra Club responded with vigorous opposition, including advertisements in national newspapers and the publications of books containing magnificent nature photography of the inner canyon.[2]

Paradoxically, recreational interest in the Colorado River of Grand Canyon was spurred by the construction of the Glen Canyon Dam upstream in 1963. The building of Glen Canyon Dam was passionately, but unsuccessfully, fought by conservationists, a battle that also attracted strong attention to the river/canyon ecosystem as a place of extraordinary beauty.[3] The paradox lies in the effect that the dam has had upon the Colorado River downstream in the Grand Canyon. The river had always been a raging torrent in places, but in its unaltered state it was heavily charged with mud and sand, "too thick to drink and too thin to plow," as the saying went. Once Glen Canyon Dam was built, the Colorado became a clear, cold river, dependent on releases from the dam, but still exciting and far more attractive for recreation than it had been in its natural state.

There was also a technological development. In place of the handcrafted rafts on which John Wesley Powell had first explored the river in 1869[4] or the hazardous canoe or kayak, large inflated rubber rafts, extremely sturdy and stable and capable of oar (or motor) navigation by a skilled boatman, were adapted from military models. Travel down these rivers became a practical possibility for large numbers of people.

The possibilities of the wild river trip were first perceived by conservation organizations, which added a few such opportunities for their members to their annual mélange of hiking, horseback, and pack trips in the backcountry. Their goal, of course, was to build a constituency of aware citizens to help preserve these rivers from development. But what they saw, others saw as well. River rafting companies began to spring up in substantial numbers. The companies hired and trained boatmen, put together package trips more or less fully outfitted for anywhere from a few days to several weeks, and advertised widely in magazines. Not surprisingly, a clientele developed, and soon the wild river trip was a familiar vacation possibility. In 1957, hardly anyone boated down the Colorado in the Grand Canyon; it was almost as unknown as it had been in Powell's day nearly a century before. By 1967, two thousand people made the trip, and by 1972, the number had increased to fifteen thousand and was still growing.[5] The next year the National Park Service froze total usage at the 1972 level;[6] the remote and mysterious canyon corridor was showing the usual signs of visitor overuse and abuse. By this time there were twenty-one commercial companies offering river trips through the Grand Canyon.[7]

Having put a limit on use at a time when river running was

becoming more and more popular, the Park Service found itself in the middle of a familiar dilemma. At one extreme were those who claimed that any form of rationing denied a right of access to many who wanted it. The Park Service's judgment about the capacity of the Canyon was challenged, and certainly capacity *is* a matter of judgment. There is little doubt that more than fifteen thousand annual users could have been found to take the trip, and enjoy it, even though there was a greater density of people, much more physical impact on the ecosystem, and a significant decline in what the Park Service viewed as an appropriate wilderness-type experience.

At the other extreme were those who preferred to have no commercial river trips offered at all, or who thought that the 1972 levels of use were grossly excessive. Just about every intermediate position was also represented. There were claims that private users should be given priority over those who went as clients of commercial companies; that the allocation the Park Service made between private and commercial users was unbalanced (at first the allocation was based on the 1972 usage by each group—frozen at 8 percent private and 92 percent commercial—and then a recommendation was made to change the ratio to 30 percent and 70 percent); and some people argued that while oar-powered boats should be allowed, motorized rafts should be prohibited.[8]

Several of these complaints led to lawsuits, each of which the Park Service won on the ground that it has broad discretion to make management judgments and that there is nothing wrong with a plan that tries—however imperfectly—to accommodate the various demands on the scarce resources of the national park system.[9]

Nonetheless, the Park Service set about trying to make a new, and more considered, management plan for river use within the park. In the usual fashion it solicited public opinion, gathered facts, held a series of public hearings around the country, and published a management plan.[10] One need not read very deeply between the lines to sense that the Park Service saw itself as caught in the middle of an impossible dilemma: too many people, too scarce resources, too much conflict among the various constituencies, and thus inevitably a solution destined to dissatisfy nearly everyone to some degree and to satisfy no one fully.

Plainly the Park Service can't satisfy everyone with only one Grand Canyon and one Colorado River, as the wide range of differing, and conflicting, views expressed at the hearings made eminently clear. It was impossible to sit through such a minidrama and not feel compassion for the bureaucratic Solomon who had to divide that baby. But the task of the Park Service at Grand Canyon need not be to satisfy everyone.

If the Park Service was given a mandate to pursue the policies suggested here, its response would be something like this: First, as an ecosystem of unusually high quality, the river within the Canyon would be reserved from conventional tourism. Access would be restricted to minimize modification of its natural ecosystem—as measured by standards such as water quality, landform protection, and maintenance of wildlife and flora; and the numbers of users would be limited so as to leave them free to set their own agenda within broad limits—to go where they want without being displaced or rushed on by other groups, to set their own pace and to linger, if they wish, so that they can experience the canyon intensely.[11]

To implement such policy, it would first be necessary to unbundle other kinds of demands that are mixed together under the management practices that presently exist. The first of these is the demand for a resort-type vacation experience, the prepackaged, tightly scheduled, and managed vacation in which every service is provided. It is clear from a reading of the brochures of commercial river-running companies that a considerable, though by no means sole, effort is made to stimulate demand for river trips from such a clientele.[12]

The most revealing parts of these brochures are the sections usually entitled "People Ask Us . . . ," or "Questions and Answers. . . ."[13] One such question is "Am I expected to work on this trip?", to which the answer is "No, please use your time on the river to hike, swim, fish, socialize or just loaf. Our trained guides will handle the rafts, pack and unpack the rafts, and prepare all meals."[14] Another brochure says "Q: What about Snakes? A: Snakes are very rarely seen. . . . They generally stay away from the camping sites . . . they don't like us any better than we like them!"[15] Or, "we take portable sanitary facilities which are as clean, comfortable, and convenient as your facilities at home . . . almost identical to the toilet units used by the airlines."[16] And "meals are expertly prepared by our guides."[17]

These, of course, are reasonable enough questions to ask of a resort manager, but they are obviously questions designed for a person whose interest in coming on the river is limited to an experience that will provide the essential qualities of a resort vacation. As with the ski resorts discussed earlier, this is a clientele whom the entrepreneur wants to attract, and whom he may need to attract to build a profitable volume of business. (Fees are equivalent to the prices in a luxury hotel.[18]) But it is not the sort of clientele that the Park Service needs to stimulate for the Colorado River. Those who want a resort experience need not be encouraged to take it on the river.

A second category of visitors are those who would simply like to raft down an exciting river. This is a form of recreational demand that should be met and public policy cannot ignore it. The quantum of this demand could be more acutely measured in the absence of commercial companies on the public lands setting out to stimulate it. Public land managers should stand ready to provide areas for those interested in river rafting, identifying some places that have good qualities for water use but are not otherwise the most pristine ecocystems. Those places can be devoted to rafting at the higher densities of use that are acceptable to many users,[19] and with a considerable range of services and planned agendas. Only as such places reach their capacities for acceptable use by the clientele, following a thorough search for alternatives, need places like the Colorado River in Grand Canyon even be considered for such use.[20]

In what would the use for which the canyon was reserved consist? Certainly it need not be limited solely to those who are able to navigate the river themselves, for this is a very small number indeed. The inner Canyon stretch of the river should, however, be limited to those who are willing to make their own schedule, to encounter snakes, and to prepare their own meals. There is no reason to prohibit people from engaging a boatman and hiring a boat, just as a mountaineer or fisherman in an unfamiliar place might engage a guide. The Park Service itself can provide interpretive trips on the river just as it provides extended interpretive programs elsewhere.[21]

Obviously there is no theoretical difference between hiring a boat and guide, and a commercial outfitter running a trip and calling himself a guide for hire. Plainly some of the commercial river-running trips are designed to do no more than provide guidance and interpretive services. The problem is not one of labeling; it is a practical problem of unbundling various kinds of demands. Under the present practice, with a plethora of concessioners offering a wide variety of services, and with strong economic incentives to stimulate additional clientele, the system works to bundle together as much as possible of what should be separate. Under the approach suggested here, the emphasis would be on the maximum possible separation.

With separation of goals in place, potential visitors would have a range of choice and a degree of clarity about what they were choosing. If the principal concern is for a package of resort-type services, visitors would know where to look, and would know clearly that those expectations would not be met in most national parks or in other high quality public land areas. Most importantly, they would know that there is a special kind of experience available to them if they go to certain places on public land. What is more, public land managers would be free to emphasize and to articulate clearly the elements of that experience, and even to stimulate demand for it. It would be possible even for novices to go down the river in Grand Canyon, but only if they were willing to do more than lie back and have someone serve them as if they were visitors at Grand Hotel, rather than Grand Canyon.

To speak of unbundling demand leaves out of consideration the individual in whom various desires are themselves mixed. Certainly it is possible to seek the experience that river running or skiing offers and, simultaneously, to be attracted by safari or resort services. I have already suggested the managerial reasons that militate in favor of separating these different desires, but is there any good reason why we should want to separate them within ourselves, and thereby to support public policies that induce us to make this separation? I think there is, and, indeed, that making such a separation lies very close to the heart of the reasons for having a distinctive park policy at all.

The issue is strikingly illustrated by the advertising literature of commercial concessioners who offer trips down wild rivers. There is in such advertising an associational quality that plays strongly upon our capacity for self-deception. To recognize that we all have competing desires to be amused and to strike out on our own is to be aware of powerful ambivalence within. To confound these competing strains so as to purport to serve longings for independence in the form of packaged entertainment is to dull the very qualities of personal engagement that give the river experience authenticity. It is rather like reading about great books so as to seem knowledgeable rather than reading the books themselves, or longing for adventure without giving up the desire for security. There is a pathetic quality in this familiar duality, and the question is whether we are ready to resist it.

At its extreme there is the commercial readiness to take an idea full of one kind of associational quality—an idea like wilderness—and to deprive it in practice of all the authentic quality that generates the association, to tame it, as Thoreau would have said. Outdoor Resorts of America, a company that has been described as being to campgrounds what Disneyland is to amusement parks, recently announced a new facility in North Carolina's Blue Ridge Mountains that would "provide campers with nearly unlimited resort amenities in a spectacular wilderness setting." Its "back-to-nature features" include stocked fishing lakes, a hunting preserve, four-wheel-drive trails, an indoor pistol range, lighted tennis courts, swimming pools, miniature golf, playground, bar, health club, live entertainment, cartoons and movies, and much, much more. The description ends by noting that "at the core of all this *wilderness and luxury* are 800 campsites . . . equipped with a paved drive, full hookups and a wood deck."[22]

Of course Outdoor Resorts of America is a private enterprise, and it can sell any kind of dreams it wishes. Whether we are well-advised to encourage such fantasies in ourselves and in the structure of public land policy is quite another matter.

The Colorado is, after all, the river that the great explorer John Wesley Powell navigated at great peril and with great intrepidity to come back a hero.[23] To invoke Powell's achievement is to suggest associations with the personal qualities of courage, independence, and self-reliance that mirror some of our most strongly felt aspirations. The wild river trip promoters play explicitly on this theme. "Would you like to retrace the journeys of some of the great explorers of history, men like Lewis and Clark or John Wesley Powell?" one brochure asks.[24] "The thrills and excitement are much as they were when Major John Wesley Powell made his historic exploration . . . ," another asserts.[25] Yet these same companies repeatedly emphasize that hired men will do the work, that the guest can lie back and relax, enjoying deluxe meals in snake-free and virtually risk-free surroundings: "For the most part, the guides do the work, while you simply enjoy the scenery. . . ."[26]

The river of Powell's 1869 is, in one sense, irrevocably gone. No one can hope to reenact fully the challenge of an original exploration and come home—five or fifteen days later in the 1980s—imbued with greatness.[27] But this was in general the case even when Powell was exploring, or some real Leatherstocking was at large in the untrammeled American wilderness. The vast majority of people were at home living lives of conventional placidity.

The practical choice has always been between an effort, within our means, to build the qualities of character such figures preeminently embodied, or simply to identify with them by some form of passive association, turning aspiration into a pale illusion. This is one of the great divides in recreational choice.

The failure to cross this divide is not limited to commercial entrepreneurs. Federal land managers themselves at times slip into serving the very illusions that a clearly articulated dual policy ought to eschew. The *United States Forest Service Manual,* attempting to explain how it develops various kinds of campsites, provides an instructive example.

Within the manual is an exhibit setting out in graphic form what purports to be a purely descriptive explanation of the different kinds of national forest camp and picnic sites, and the recreational experiences associated with them.[28] Thus, along one column there appears a five-fold classification of types of site development, ranging from primitive to modern; and along a parallel column a description of the associated "recreation experiences" engendered by them. The first column, for example, describes a primitive site as one with very few facilities, a genuinely natural setting with native foliage intact, and only such improvements as are required for the physical protection of the site (perhaps a ring of rocks indicating a campfire site, or a small cleared area suggesting that campers should stay back a certain distance from a lake). The categories describe increasingly developed sites, culminating in the "modern" area, which will have access by high-speed roads, replacement of a natural forest environment with clipped shrubs and mown grass lawns, and facilities such as flush toilets, showers, bath houses, laundry facilities, and electrical hookups.

The idea, of course, is that the Forest Service is providing a range of opportunities for everyone from the wilderness backpacker to the driver of a fully electrified recreation vehicle. All this is routine enough. What makes the exhibit fascinating is the way in which it describes the needs of the various users for whom these facilities are provided. A more conventional document of this sort would simply say that some people want electric hookups and showers, and others want hiking trails in the woods. But this exhibit comes about as close as is possible to saying that people want illusions that provide the comfort of familiar services while suggesting self-reliant adventure, and that the Forest Service is calculatingly giving them exactly that.

Consider the recreation experience of the most developed modern site. Such a place, the exhibit notes, is designed to "satisfy the urbanite's need for compensating experiences and relative solitude" (precisely the need that induces people to leave the city and seek out a natural environment); but, it continues, it is "obvious to user that he is in secure situation where ample provision is made for his personal comfort and he will not be called upon to use undeveloped skills." In short, to the visitor who can barely, but only barely, sense a need to break out of the managed pattern of urban life, such a site will give the sense that it is possible to do so without in any respect suggesting the burden that is necessary to make that break. Indeed, in such a site "regimentation of users is obvious"—and comforting.

This leads to the next category, which is designated secondary modern. Here the "contrast to daily living routines is moderate. Invites marked sense of security." As one moves up the ladder toward the more primitive site, managerial control becomes increasingly less visible, but not less real. The intermediate site continues to be significantly developed, but native materials are used to provide the sense of a natural setting, and control of patterns of movement is "inconspicuous." The goal is to provide enough "control and regimentation" for the safety of the user, and to make them "obvious enough to afford a sense of security but subtle enough to leave the taste of adventure." By the time one gets to the two least developed categories, primitive and secondary primitive, it is clear that the taste of adventure has some reality to it, but even in the primitive category the unseen presence of the Forest Service managers is powerful. The camper at a primitive site "senses no regimentation," but what he senses is not exactly what he gets, for while there are only minimal controls, these "minimum controls are subtle." There is just "no obvious means [of] regimentation."

Perhaps this description makes the Forest Service's plan seem rather sinister. In fact there is nothing sinister about the goal. The Forest Service—like the Park Service at Grand Canyon—has been providing a rather wide range of outdoor recreation experiences graduated to the range of demands it perceives in its clientele.[29] They sense quite accurately a desire for "controls," "regimentation," and "security," and at the same time a demand for "a taste of adventure," "solitude," and "testing of skills."

If these competing intuitions were unbundled and presented separately as choices to be faced we would be well on the way to a mature system of public recreation attuned to the ambivalence within.[30] Instead we are offered illusions.[31]

8

Conclusion

It must seem curious that a book about the national parks talks so little about nature for its own sake, and may even seem to denigrate ecosystem preservation as central to the mission of the parks. My only explanation is that most conflict over national park policy does not really turn on whether we ought to have nature reserves (for that is widely agreed), but on the uses that people will make of those places—which is neither a subject of general agreement nor capable of resolution by reference to ecological principles. The preservationists are really moralists at heart, and people are very much at the center of their concerns. They encourage people to immerse themselves in natural settings and to behave there in certain ways, because they believe such behavior is redeeming.

Moreover, the preservationists do not merely aspire to persuade individuals how to conduct their personal lives. With the exception of Thoreau, who predated the national park era, they have directed their prescriptions to government. The parks are, after all, public institutions which belong to everyone, not just to wilderness hikers. The weight of the preservationist view, therefore, turns not only on its persuasiveness for the individual as such, but also on its ability to garner the support—or at least the tolerance—of citizens in a democratic society to bring the preservationist vision into operation as official policy. It is not enough to accept the preservationists simply as a minority, speaking for a minority, however impressive. For that reason I have described them as secular prophets, preaching a message of secular salvation. I have attempted to articulate their views as a public philosophy, rather than treating them merely as spokesmen for an avocation of nature appreciation, because the claims they make on government oblige them to bear the weightier burden.

This is not to say that what they preach cannot be rejected as merely a matter of taste, of elitist sentiment or as yet another reworking of pastoral sentimentalism. It is, however, to admit that their desire to dominate a public policy for public parks cannot prevail if their message is taken in so limited a compass. If they cannot persuade a majority that the country needs national parks of the kind they propose, much as it needs public schools and libraries, then the role they have long sought to play in the governmental process cannot be sustained. The claim is bold, and it has often been concealed in a pastiche of argument for scientific protection of nature, minority rights, and sentimental rhetoric. I have tried to isolate and make explicit the political claim, as it relates to the fashioning of public policy, and leave it to sail or sink on that basis.

It may also seem curious that I have put the preservationists into the foreground, rather than the Congress or the National Park Service. Of course Congress has the power to be paternalistic if it wishes, and it often is. It thinks a lot of things are good for us, from free trade and a nuclear defense system to public statuary and space exploration. But no unkindness is intended by the observation that Congress doesn't really think at all. At best it responds to the ideas that thinkers put before it, considers the merits of those thoughts, tests them against its sense of the larger themes that give American society coherence, and asks whether the majority will find them attractive or tolerable. The fundamental question then—and the question I have tried to address here—is whether the ideas of nature preservationists meet these tests. If they do, Congress will ultimately reflect them.

The National Park Service, and other bureaucracies that manage nature reserves, are also basically reflective institutions. Strictly speaking, they enforce the rules Congress makes, doing what they are told. But no administrative agency is in fact so mechanical in its operation. It has its own sense of mission, an internal conception of what it ought to be doing, and that sense of mission also harks back to what thinkers have persuaded it, institutionally, to believe. If the Park Service is basically dominated by the ideology of the preservationists, it will act in certain ways, given the opportunity. If, on the other hand, it has come to believe in the commodity-view of the parks, it will behave quite differently. Thus, again, the capacity of the preservationist view to persuade is the essential issue.

At the same time, no bureaucracy behaves simply according to its own sense of mission. It lives in a political milieu, with constituencies of users and neighbors who impose strong, and at times irresistable, pressures on it. What the general public believes about the appropriate mission of the national parks is also essential. If the preservationist is to prevail, he must gain at least the passive support of the public, which will indirectly be felt by the Park Service in the decisions it makes in day to day management.

For these reasons, the preceding pages have been devoted to what preservationists *think* about the national parks, rather than to park history, assessments of popular demand, or the rules officials have, in fact, made. The preservationist message is addressed to three audiences simultaneously: the Congress, the National Park Service, and the general public.

To Congress it says don't try to make the national parks all things to all people in every location. Do use the public lands to serve conventional recreational preferences, but save some places explicitly for what has been called—lacking any fully satisfactory term—reflective or contemplative recreation. Indeed, try to encourage more of such recreation, and for that reason try to accommodate conventional demands, as much as you can, at other places. Moreover, make some effort to discourage the use of public lands for those forms of recreation that are the most consumptive of the resources, and that rest principally on the inclination toward power and dominion. In so doing, you will both stretch the capacity of our physical resources to meet public recreational needs, and also play a minimally coercive role in giving leadership to a culture value that is worthy of support.

To the Park Service, the message is that your traditional inclination to associate yourselves with the preservationist tradition should be encouraged. Nothing in that tradition intrudes upon the basic values of a democratic society that you are obliged to uphold. Hold fast to the position that park visitors have a duty to themselves to unbundle their various recreational desires: force the duty to choose on them and resist pressures to deprive the parks of authenticity under such labels as "threshold wilderness experiences."

Most importantly, however, if you do commit yourself to the preservationist view, keep in mind that its implementation is most important in those places where the great bulk of visitors find themselves, places like Yosemite Valley. Do not, therefore, write off such places as irrevocably urbanized, and do not congratulate yourselves simply because the bulk of acreage in the parks is still undeveloped wilderness. Pursue, but even more forcefully, your emerging plans for the reduction of development in places like Yosemite Valley. Pursue also the distinctive role carved out for urban parks that is reflected, at least in part, in the early planning for New York's Gateway East.

Another important message may be drawn from the Olympic Park controversy described earlier. It is that much as public pressures may seem to restrict your scope of action, keep in mind that a bold affirmative strategy is the most effective way to keep those pressures at bay. You must yourselves seek out alternative sites for the users who claim that only parklands can adequately serve conventional tourism. You cannot simply reject pressures without offering attractive substitutes. You need to know much more about what resources are available elsewhere. You can't build a wall around the parks and close your eyes to what goes on outside them. You will also have to be more skillful in dealing with private entrepreneurs and concessioners who consciously build demands for the use of parklands that are extremely difficult to deal with after the fact. If you had been more foresighted in the management of the Colorado River in Grand Canyon, for example, and had looked ahead before some twenty commercial operators began generating customers for this remote and fragile place, you would not have had so difficult a time in managing the river as you would like.

You can also affect the popular pressures you feel by affirmatively encouraging opportunities for less urbanized recreation outside the national parks. Ultimately the parks will reflect the kinds of recreation habits most people have. There are many ways in which people can be encouraged to use their leisure time in a slower-paced, less energy-consuming, and more intensive fashion. An old tradition in vacation styles, where people went into a community and lived with local people, learning to savor the indigenous style and pace of the area, could usefully be revived; and doing so would help develop a new constituency amenable to the preservationist prescription for the parks.

The traditional practice of government has been to promote tourism by building new, high-speed roads in an area and encouraging the construction of modern highway motels. Indeed, one sees many such developments near the national parks themselves, and their style of tourism powerfully affects the parks. The result of such government programs, however, is often neither to advantage the local population, which finds itself with dead-end service jobs in motels and stores, nor to give the tourist any distinctive sense of the area he is visiting. He stays in a conventional motel, sees only what is seen by car, and shops in souvenir shops and curio stands that have grown up solely to serve him.

There are many places where a different sort of strategy could pay long-lasting dividends. The fascinating Indian and Spanish communities of the Southwest exemplify one such an opportunity. By immersing himself in the regional culture, the visitor could experience the unique qualities the area possesses, and local people would have the opportunity to benefit more directly from tourist traffic. Rather than being encouraged to abandon their culture to become busboys and waiters, they would have an incentive to maintain the distinctive qualities of their own community.

Successful examples of such efforts already exist. There are commercial enterprises that arrange for American families to spend vacations living with families in the French countryside.[1] The host family, though it is paid for the board and lodging it provides, is just that—a host—offering hospitality and an opportunity to see something of the region from the inside, while maintaining its own dignity, status, and distinctiveness, rather than abandoning it to the international motel style of foreign tourism.

Such arrangements would not directly affect the national parks, for of course there are very few human settlements in the parks themselves. But they could have a profound indirect effect. They would encourage a more deliberate, more probing, style of tourism, with less incentive to change existing communities (both natural and human) to meet the visitor's preconception, and instead encourage the visitor to adapt to the setting of the place visited. Changed attitudes about recreation in the large and in the long run are fundamentally what will determine the future of the parks, and an imaginative National Park Service bureaucracy will have to take the large view if it is to play a significant role in that future. Efforts directed to stimulating enjoyment of an area's distinctive character will ultimately generate appreciation for the features that give a place what René Dubos calls its "genius,"[2] its authenticity of character, whether it is a languid desert, a remote mountain community of villagers, or a bleak intersection of land and seascape.

To those for whom wilderness values and the symbolic message of the parks has never been of more than peripheral importance, this book asks principally for tolerance: a willingness to entertain the suggestion that the parks are more valuable as artifacts of culture than as commodity resources; a willingness to try a new departure in the use of leisure more demanding than conventional recreation; a sympathetic ear tuned to the claim for self-paternalism.

Finally, to the preservationists themselves, in whose ranks I include myself, the message is that the parks are not self-justifying.

Your vision is not necessarily one that will commend itself to the majority. It rests on a set of moral and aesthetic attitudes whose force is not strengthened either by contemptuous disdain of those who question your conception of what a national park should be, or by taking refuge in claims of ecological necessity. Tolerance is required on all sides, along with a certain modesty. "I have gone a-fishing while others were struggling and groaning and losing their souls in the great social or political or business maelstrom," the nature writer John Burroughs observed late in his life. "I know too I have gone a-fishing while others have labored in the slums and given their lives to the betterment of their fellows. But I have been a good fisherman, and I should have made a poor reformer. . . . My strength is my calm, my serenity."[3]

A Policy Statement: The Meaning of National Parks Today

I. *The parks are places where recreation reflects the aspirations of a free and independent people*

They are places where no one else prepares entertainment for the visitor, predetermines his responses, or tells him what to do. In a national park the visitor is on his own, setting an agenda for himself, discovering what is interesting, going at his own pace. The parks provide a contrast to the familiar situation in which we are bored unless someone tells us how to fill our time.

The parks are places that have not been tamed, contemporary symbols for men and women who are themselves ready to resist being tamed into passivity. The meaning of national parks for us grows as much out of the modern literature of inner freedom and its fragility as it does out of traditional nature writing. "I tell thee," Dostoevski says in *The Brothers Karamazov,* "that man is tormented by no greater anxiety than to find someone quickly to whom he can hand over that gift of freedom with which the ill-fated creature is born."

II. *The parks are an object lesson for a world of limited resources*

In the national parks the visitor learns that satisfaction is not correlated to the rate at which he expends resources, but that just the opposite is true. The parks promote intensive *experience,* rather than intensive *use.* The more one knows, searches, and understands, the greater the interest and satisfaction of the park experience.

To a very casual visitor, even the stupendous quality of a Grand Canyon is soon boring; he yearns for "something to do." The more the visitor knows about the setting, however, the greater its capacity to interest and engage him. He cannot exhaust its interest in a lifetime. In the same way, the more knowledgeable and engaged the visitor, the less he wants or needs to pass through the parks quickly or at high speed. The quantity of resources the visitor needs to consume shrinks as he discovers the secret of intensiveness of experience, and his capacity for intense satisfaction depends on what is in his own head. This of course is what Thoreau meant when in his famous essay "Walking" he said:

> My vicinity affords many good walks; and though I have walked almost every day . . . I have not yet exhausted them. . . . The limits of an afternoon walk . . . will never become quite familiar to you.

The parks perform their function without being used up at all. We do not increase our enjoyment of an alpine meadow by picking its flowers, but by leaving them where they are. The more we understand that they are part of a larger system the more we appreciate them in their setting; and in their setting—rather than perceived as things to possess or to use up—they are inexhaustible.

III. *The parks are great laboratories of successful natural communities*
We look at nature with awe and wonderment: Trees that have survived for millenia; a profusion of flowers in the seeming sterility of the desert; predator and prey living in equilibrium; undiminished productivity and reproduction, year after year, century upon century. These marvels intrigue us, but nature is also a model of many things we seek in human communities. We value continuity, stability, and sustenance. And we see in nature attainment of those goals through adaptation, sustained productivity, diversity, and evolutionary change.

Ideas are perhaps the scarcest of all resources, and nature is a cornucopia of ideas in a vast laboratory setting. With a discerning eye, one can see in any park a multitude of examples of efficiency and adaptation—in architecture, in food production and gathering, in resistance to disease, in procreation, and energy use—all of which have counterparts in human society. Our interest in preserving natural systems is not merely sentimental; it rests on preservation of nothing less than an enormous knowledge base that we have no capacity to replicate. To some these are merely practical benefits; to some, they suggest ethical imperatives. Whatever our final characterization, nature provides an unequaled storehouse of material for human contemplation.

IV. *The parks are living memorials of human history on the American continent*

For the most part, the national parks demonstrate the continuity of natural history measured over millenia. The less dramatic span of human settlements is an equally essential part of that history, and the national park system is a richly endowed showcase of our history as a people. Here too adaptation and succession, struggle and continuity, diversity and change are revealed. The settlements of Native American peoples at places like Mesa Verde and Bandelier National Monument; the great sites associated with the American revolution, such as Boston National Historical Park and the principal Civil War battlefields; the communities of early settlers—Cade's Cove in the Great Smoky Mountains and the mining at Death Valley; the Robert E. Lee home and the Booker T. Washington birthplace.

These places are essential to the aspirations of a free people, for without our history we are at large and vulnerable in the present. In *1984,* George Orwell's great novel of freedom and its loss, one of the most poignant scenes is that where the hero Winston searches for a perspective against which to measure the life in which an all-powerful state had immersed him. But the past had been obliterated: "He tried to squeeze out some childhood memory that should tell him whether London had always been quite like this . . . it was no use . . . nothing remained. . . ."

Notes

Introduction

1. The term *preservationist* is often used in an uncomplimentary way as a synonym for *extremist*. John McPhee, *Encounters with the Archdruid* (New York: Farrar, Straus and Giroux, 1971), p. 95. Here it has no such connotation. By preservationists, I mean those whose inclinations are to retain parklands largely (though not absolutely) as natural areas, without industrialization, commercialized recreation, or urban influences. There is no official preservationist position, and obviously no unanimity of view on any controversial question. Among the organizations that speak most consistently for such views are the National Parks and Conservation Association, the Wilderness Society, the Sierra Club, and Friends of the Earth. The bibliography at the end of this book surveys, though with no attempt at comprehensiveness, major influences in the preservationist literature.

   It is my thesis that, however unself-consciously, those who speak for essentially undeveloped parks adhere to a set of generally consistent views. One major purpose of this book is to draw together and articulate those views as a coherent ideology for the parks.
2. While this is a book about national parks, by no means all the lands of national park quality are thus designated by the Congress. American public land history is a tangled web of confusing categories. Much of our high quality wilderness is within the national forests, administered by the Department of Agriculture, rather than by the Interior Department's National Park Service. The greatest acreage of all is administered by the Interior Department's Bureau of Land Management (BLM). And a good deal of

superlative parkland is owned and managed by the states, or even by local governments.

Even the official "national park system" is a mélange of national parks, national monuments, national recreation areas, and national lakeshores and seashores, as well as numerous historic sites and other miscellany. See *Index of the National Park System* (Washington, D.C.: Government Printing Office, 1979). Each of these categories is managed under different legislative mandates.

My concern is not with the numerous distinctions Congress has made, but with the general question of how we ought to want to use our high-quality natural areas held in public ownership. Though I use the term national parks throughout, much of what I say is applicable to certain national forest and BLM lands, to state parks, and to a range of lands within the national park system, however designated.

Chapter 1

1. The standard history of the national parks is John Ise, *Our National Park Policy: A Critical History* (Baltimore: Johns Hopkins University Press, 1961). See also, Alfred Runte, *National Parks: The American Experience* (Lincoln: University of Nebraska Press, 1979); Hans Huth, *Nature and the American: Three Centuries of Changing Attitudes* (Lincoln: University of Nebraska Press, 1972); Roderick Nash, *Wilderness and the American Mind* (New Haven: Yale University Press, 1973); Joseph L. Sax, "America's National Parks: Their Principles, Purposes and Prospects," *Natural History* 85, no. 8 (October, 1976):57.
2. See Huth, *Nature and the American,* chap. 4.
3. Travelers' books and artistic work had, however, made western scenery familiar and popular. Huth, *Nature and the American,* chap. 8.
4. Act of June 30, 1864, 13 Stat. 325, ch. 184, §§ 1,2. Huth, *Nature and the American,* chap. 9.
5. *Hutchings Illustrated California Magazine* 4, no. 4 (October, 1859):145.
6. For the early history of Yellowstone, see Aubrey L. Haines, *Yellowstone National Park: Its Exploration and Establishment* (Washington, D.C.: Government Printing Office, 1974).
7. Hans Huth, "Yosemite: The Story of an Idea," *Sierra Club Bulletin,* 33, no. 3 (March, 1948):67.
8. Huth, *Nature and the American,* p. 223 n.12.

9. See note 4 above.
10. Act of March 1, 1872, 17 Stat. 32, ch. 24, § 1.
11. The history of the early parks is described in Ise, *Our National Park Policy.*
12. Act of August 25, 1916, 39 Stat. 535, ch. 408, § 1. Donald C. Swain, "The Passage of the National Park Service Act of 1916," *Wisconsin Magazine of History* 50 (1966):4, 9–11.
13. H. Duane Hampton, *How the U.S. Cavalry Saved Our National Parks* (Bloomington: Indiana University Press, 1971).
14. See Runte, *National Parks;* also Alfred Runte, "The National Park Idea: Origins and Paradox of American Experience," *Journal of Forest History* 21, no. 2 (April, 1977):64.
15. See note 7 above.
16. Nash, *Wilderness and the American Mind,* pp. 130–32.
17. Huth, "Yosemite," p. 63.
18. Hampton, *How the U.S. Cavalry Saved Our National Parks,* pp. 40–41.
19. *The Maine Woods,* ed. Joseph J. Moldenhauer (Princeton: Princeton University Press, 1972), p. 121.
20. Carolyn de Vries, *Grand and Ancient Forest: The Story of Andrew P. Hill and Big Basin Redwood State Park* (Fresno, Calif.: Valley Publishers, 1978), p. 28. This attitude was doubtless a reaction to the excesses in the "reformation" of the wilderness that had seemed appropriate to earlier American thinkers. See Cecelia Tichi, *New World, New Earth: Environmental Reform in American Literature from the Puritans Through Whitman* (New Haven: Yale University Press, 1979).
21. Paul W. Gates and Robert W. Swenson, *History of Public Land Law Development* (Washington, D.C.: Government Printing Office, 1968), chaps. 15, 22.
22. Erik Nilsson, *Rocky Mountain National Park Trail Guide* (Mountain View, Calif.: World Publications, 1978), pp. 6–9.
23. Runte, *National Parks,* pp. 77, 93; Nash, *Wilderness and the American Mind,* p. 68; Haines, *Yellowstone National Park,* pp. 126, 128; Huth, *Nature and the American,* pp. 36, 38, 49.
24. Ise, *Our National Park Policy,* p. 196. Robert Shankland, *Steve Mather of the National Parks* (New York: Alfred A. Knopf, 1951), p. 145.
25. Runte, *National Parks,* p. 48.
26. Redwood National Park was not established until 1968. Act of

October 2, 1968, 82 Stat. 931, Public Law 90–545, § 1. Congress enlarged the park in 1978. Act of March 27, 1978, 92 Stat. 163, Public Law 95–250, Title I, § 101(a)(1). See "The Tragedy of Redwood National Park," *Natural Resources Defense Council Newsletter* 6, no. 4, (July/August, 1977):1.

27. On the Hetch Hetchy battle, see Holway R. Jones, *John Muir and the Sierra Club: The Battle for Yosemite* (San Francisco: Sierra Club, 1965).
28. Ise, *Our National Park Policy,* pp. 307–17.
29. There were, of course, always concessions to commercial interests. Crater Lake National Park was opened to mining in 1902 (Act of May 22, 1902, 32 Stat. 202), and reclamation projects were permitted in both Rocky Mountain and Glacier National parks (Act of January 26, 1915, 38 Stat. 798; Act of May 11, 1910, 36 Stat. 354).
30. Harold K. Steen, *The U.S. Forest Service: A History* (Seattle: University of Washington Press, 1976), p. 124.
31. Ibid., pp. 118–22.
32. Shankland, *Steve Mather of the National Parks,* p. 97.
33. Earl Pomeroy, *In Search of the Golden West: The Tourist in Western America* (New York: Alfred A. Knopf, 1957), pp. 151–52. Almost from the beginning, Yosemite Valley has suffered from overdevelopment of various kinds. Shirley Sargent, *Yosemite and Its Innkeepers* (Yosemite, Calif.: Flying Spur Press, 1975), pp. 87, 89, 96, 116–17. Speaking of early tourists in Yosemite, John Muir wrote disparagingly: "They climb sprawlingly to their saddles like overgrown frogs, ride up the Valley with about as much emotion as the horses they ride upon and, comfortable when they have 'done it all', . . . long for the safety and flatness of their proper homes." John Muir, *Letters to a Friend* (Boston: Houghton Mifflin Co., 1915), pp. 80–81.
34. "The service thus established shall promote and regulate the use of the . . . national parks . . . by such means and measures as conform to the fundamental purpose of the said parks . . . which purpose is to conserve the scenery and the natural and historic objects and the wild life therein and to provide for the enjoyment of the same in such manner and by such means as will leave them unimpaired for the enjoyment of future generations." Act of August 25, 1916, 39 Stat. 535, ch. 408, § 1.
35. See, e.g., U.S. Department of the Interior, National Park Service, *Draft General Management Plan, Yosemite National Park, California*

(August, 1978), revised January, 1980. Compare with the (rejected) *Yosemite National Park Preliminary Draft Master Plan* (August 12, 1974). See Ise, *Our National Park Policy,* p. 437.

36. U.S., Congress, House, *National Park Service Planning and Concession Operations: Joint Hearing Before Certain Subcommittees of the Committee on Government Operations and the Permanent Select Committee on Small Business,* 93d Cong., 2d sess., December 20, 1974. Of course there were always some large concessioners, such as the railroads. See text at note 8 above.
37. E.g., Fred B. Eisemen, Jr., "Who Runs the Grand Canyon?", *Natural History* 87, no. 3 (March, 1978): 83–93; Liz Hymans, "The Flow of Wilderness," *Not Man Apart* (Friends of the Earth) 8, no. 5 (Mid-March, 1978).
38. *Desert Solitaire* (New York: Random House, Ballantine Books, 1971), p. 52. See also Edward Abbey and Philip Hyde, *Slickrock: The Canyon Country* (San Francisco: Sierra Club, 1971), p. 71. A somewhat more tolerant view is revealed in "The Winnebago Tribe," in Edward Abbey, *Abbey's Road* (New York: E. P. Dutton, 1979), pp. 142–44.
39. See chapter 7.

Chapter 2

1. *A Sand County Almanac with Other Essays on Conservation from Round River* (New York: Oxford University Press, 1966), p. 196.
2. T. H. Watkins and Dewitt Jones, *John Muir's America* (New York: Crown Publishers, 1976), p. 141.
3. Reprinted in *Landscape Architecture* 44, no. 1 (1953):17.
4. Hans Huth, "Yosemite: The Story of an Idea," *Sierra Club Bulletin* 33, no. 3 (March, 1948):66.
5. Introduction by Laura Wood Roper to Olmsted's report, *Landscape Architecture* 44, no. 1 (1953):13.
6. Ibid.
7. The report has been discussed occasionally in published sources. E.g., Holway R. Jones, *John Muir and the Sierra Club: The Battle for Yosemite* (San Francisco: Sierra Club, 1965), p. 30–32; and in Laura Wood Roper's biography, *FLO: A Biography of Frederick Law Olmsted* (Baltimore: Johns Hopkins University Press, 1973), pp. 283–87. See also U.S. Council on Environmental Quality, *Third Annual Report* (1972), p. 313. On October 12, 1979, Congress finally established a Frederick Law Olmsted National Historic Site. Public Law 96–87.

8. "The Yosemite Valley and the Mariposa Big Trees," *Landscape Architecture* 44, no. 1 (1953):17.
9. Ibid., p. 20.
10. Ibid.
11. Ibid., p. 21.
12. Ibid.
13. Ibid.
14. Ibid. Notably, Olmsted's ideas parallel the views of those who promoted the establishment of libraries, universities, and museums. William B. Ashley, "The Promotion of Museums," *Proceedings of the American Association of Museums* 7 (1913):39, 44; Frederick Rudolph, *The American College and University: A History* (New York: Alfred A. Knopf, 1962), p. 279; Sidney Ditzion, *Arsenals of a Democratic Culture* (Chicago: American Library Association, 1947), p. 153. Olmsted's view was, of course, an article of faith in nineteenth-century America; see Frank Moore, ed., *Andrew Johnson: Speeches* (Boston: Little, Brown, 1866), p. 56.
15. "The Yosemite Valley," p. 22.
16. Ibid., p. 20.
17. Charles M. Dow, *The State Reservation at Niagara: A History* (Albany: J. B. Lyon Co., 1914).
18. "The Movement for the Redemption of Niagara," *New Princeton Review* (New York, A. C. Armstrong & Sons, 1886), 1:233–45; Alfred Runte, "Beyond the Spectacular: The Niagara Falls Preservation Campaign," *New York State Historical Society Quarterly* 57 (January, 1973):30–50.
19. "Notes by Mr. Olmsted," *Special Report of the New York State Survey on the Preservation of the Scenery of Niagara Falls and Fourth Annual Report on the Triangulation of the State for the Year 1879* (Albany: Charles von Benthuysen & Sons, 1880), p. 27.
20. "Report of Messrs. Olmsted and Vaux," *Supplemental Report of the Commissioners of the State Reservation at Niagara Transmitted to the Legislature January 31, 1887* (Albany: Argus Co., 1887), p. 21. The description that follows in the text is taken from this and the 1879 report.
21. See note 19 above.
22. Ibid.
23. "The Yosemite Valley and the Mariposa Big Trees," p. 21.
24. See note 19 above.
25. "Nature," in Ralph Waldo Emerson, *Selected Prose and Poetry,* 2d

ed., ed. Reginald L. Cook, (New York: Holt, Rinehart and Winston, Rinehart Editions, 1969), p. 38.

26. "The American Scholar," ibid., p. 55.
27. See generally Raymond Williams, *The Country and the City* (New York: Oxford University Press, 1973).

Chapter 3

1. There is a related scholarly literature. E.g., Mihalyi Csikszentmihalyi, *Beyond Boredom and Anxiety* (San Francisco: Jossey-Bass, 1975); Johan Huizinga, *Homo Ludens: A Study of the Play Element in Culture* (London: Routledge & Kegan Paul, 1949); Roger Caillois, *Man, Play and Games* (New York: Free Press, 1961); Paul Weiss, *Sport: A Philosophic Inquiry* (Carbondale: Southern Illinois University Press, 1969); Allen Guttman, *From Ritual to Record: The Nature of Modern Sports* (New York: Columbia University Press, 1978).
2. Extensive bibliographies are provided in Arnold Gingrich. *The Well-Tempered Angler* (New York: Alfred A. Knopf, 1966), pp. 317–23 and *The Joys of Trout* (New York: Crown Publishers, 1973), pp. 251–64.
3. *The Joys of Trout,* p. 6.
4. *A River Never Sleeps* (New York: William Morrow & Co., 1946), p. 268.
5. (New York: Crown Publishers, 1971), p. xiii.
6. (London: Navarre Society, 1925), pp. 48–49.
7. Ibid., pp. 114, 129.
8. Ibid., p. 66.
9. Ibid., pp. 43, 51.
10. (Chicago: University of Chicago Press, 1976), p. 42.
11. *A River Never Sleeps,* p. 250.
12. Ibid., pp. 79–80.
13. See, e.g., Colin Fletcher, *The Man Who Walked Through Time* (New York: Random House, Vintage Books, 1967).
14. *A River Never Sleeps,* p. 272.
15. (New York: Charles Scribner's Sons, 1972). See also Stephen R. Kellert, "Attitudes and Characteristics of Hunters and Antihunters," in *Transactions of the 43rd North American Wildlife and Natural Resources Conference* (Washington, D.C.: Wildlife Management Institute, 1978), pp. 412–23; Paul Shepard, Jr., *The Tender Carnivore and the Sacred Game* (New York: Charles Scribner's Sons, 1973), chap. 4.

16. Gasset, *Meditations on Hunting,* pp. 75, 35.
17. "Wildlife in American Culture," in *A Sand County Almanac with Other Essays on Conservation from Round River* (New York: Oxford University Press, 1966), p. 197.
18. Ibid. This is an observation routinely made in the mountaineering literature. See Csikszentmihalyi, *Beyond Boredom and Anxiety,* pp. 48, 75; Jeremy Bernstein, *Mountain Passages* (Lincoln: University of Nebraska Press, 1978), p. 37; Geoffrey Winthrop Young, *Mountain Craft* (London: Methuen Co., 1949), p. 3.
19. The best discussion appears in the report of the United States Council on Environmental Quality, David Sheridan, "Off-Road Vehicles on Public Land" (Washington, D.C.: Government Printing Office, 1979).
20. Robert Pirsig, *Zen and the Art of Motorcycle Maintenance* (New York: Bantam Books, 1976).
21. E.g., J. Ginsberg, R. Mintz, and W. S. Walter, *The Fragile Balance: Environmental Problems of the California Desert* (Stanford: Stanford Law School Environmental Law Society, 1976), chap. 3.
22. Sheridan, "Off-Road Vehicles on Public Land," p. 2.
23. Lee Gutkind, *Bike Fever* (New York: Avon Books, 1974), pp. 211–13.
24. Sally Wimer, *The Snowmobiler's Companion* (New York: Charles Scribner's Sons, 1973), pp. x, 150.
25. A characteristic older book is Guido Rey, *The Matterhorn* (London: T. Fisher Unwin, 1907). The modern genre is illustrated by Maurice Herzog, *Annapurna* (New York: E. P. Dutton & Co., 1953). A. C. Spectorsky, ed., *The Book of the Mountains* (New York: Appleton-Century-Crofts, 1955) contains an extensive collection of mountaineering writing. See also note 18 above; and notes 26, 27, 28 below.
26. *The Maine Woods* (Princeton: Princeton University Press, 1972), p. 65.
27. Chris Bonington, *Everest the Hard Way* (New York: Random House, 1976).
28. Galen Rowell, *In the Throne Room of the Mountain Gods* (San Francisco: Sierra Club, 1977). See also Galen Rowell, *High and Wild: A Mountaineer's World* (San Francisco: Sierra Club, 1979).
29. Csikszentmihalyi, *Beyond Boredom and Anxiety,* p. 75.
30. Rowell, *In the Throne Room of the Mountain Gods,* pp. 111, 178, and Bonington, *Everest the Hard Way,* p. 14.
31. Bonington, *Everest the Hard Way,* p. 14.

32. Rowell, *In the Throne Room of the Mountain Gods,* p. 111.
33. Ibid.
34. Quoted in Rowell, *In the Throne Room of the Mountain Gods,* p. 147.
35. "A Perilous Night on Mount Shasta," from *Steep Trails,* in Edwin Way Teale, ed., *The Wilderness World of John Muir* (Boston: Houghton Mifflin Co., 1976), pp. 251–65.
36. Norman Foerster, *Nature in American Literature* (New York: Russell & Russell, 1923), p. 245.
37. Rowell, *In the Throne Room of the Mountain Gods,* pp. 169–70.
38. Ibid., p. 157.
39. Ibid., p. 110.
40. *Ascent,* 1964, pp. 23–25.
41. Ibid., p. 23.
42. Ibid., p. 24.
43. "The Climber As Visionary," *Ascent,* 1969, p. 6.
44. Ibid.
45. U.S., Department of the Interior, National Park Service, Office of Public Affairs, "Editorial Briefs," 7, no. 34 (August 21, 1979), p. 3.
46. "The Bear," in *The Portable Faulkner,* ed. Malcolm Cowley (New York: Viking Press, 1967), p. 199.
47. Ibid., p. 197.
48. In the collection *In Our Time* (New York: Charles Scribner's Sons, 1970), pp. 131–56. Another lovely Hemingway fishing scene appears in chapter 7 of *The Sun Also Rises* (New York: Charles Scribner's Sons, 1926).
49. Ibid., p. 134.
50. The background of "Big Two-Hearted River" is explained in a later Hemingway story, "A Way You'll Never Be," in Ernest Hemingway, *The Nick Adams Stories* (New York: Bantam Books, 1973), p. 135.
51. (New York: Charles Scribner's Sons, 1952). The hunt is a persistent and complex theme in Hemingway's writings, as in his life. The safari in *The Green Hills of Africa* (New York: Charles Scribner's Sons, 1963) and the exegesis of bullfighting in *Death in the Afternoon* (New York: Charles Scribner's Sons, 1960) both show a fascination with the kill characteristic of the author's own behavior. Carlos Baker, *Ernest Hemingway: A Life Story* (New York: Charles Scribner's Sons, 1969).
52. *The Old Man and the Sea,* p. 50.
53. Ibid., p. 54.
54. Johan Huizinga, *The Waning of the Middle Ages* (Garden City,

N.Y.: Doubleday, Anchor Books, 1954), p. 39. See generally Raymond Williams, *The Country and the City* (New York: Oxford University Press, 1973).

55. *Virgin Land: The American West as Symbol and Myth* (Cambridge, Mass.: Harvard University Press, 1950).
56. *Public Parks and the Enlargement of Towns* (Cambridge, Mass.: Riverside Press, 1870; reprint ed., New York: Arno Press and The New York Times, 1970), pp. 22–23.
57. E.g., see Muir's advice to businessmen in "The Gospel for July," *Sunset* 23, no. 1 (July, 1909):1. See also Joseph L. Sax, "Freedom: Voices From The Wilderness," *Environmental Law* 7 (1977):568; John Hammond, "Wilderness and Life in Cities," *Sierra: The Sierra Club Bulletin* 62, no. 4 (April, 1977):12–14.
58. *Alaska Wilderness* (Berkeley and Los Angeles: University of California Press, 1973), p. 165. See also Marshall's "The Problem of Wilderness," *Scientific Monthly* 30 (February, 1930):141–48.
59. Carl Bode, ed., *The Portable Thoreau* (New York: Viking, 1975), pp. 456–57. Note the strikingly parallel description in Rowell, *In the Throne Room of the Mountain Gods,* p. 145.
60. "The [nature] hunter deeply respects and admires the creature he hunts. This is the mysterious, ancient contradiction of the real hunter's character—that he can at once hunt the thing he loves." J. Madson and E. Kozicky, "The Hunting Ethic," *Rod and Gun* 166, no. 3 (1964):12, quoted in Kellert, "Attitudes and Characteristics of Hunters and Anti-hunters," p. 415. See note 46 above.
61. Bode, *Portable Thoreau,* p. 458.
62. Ibid., pp. 460, 458, 459. See also *The Maine Woods,* ed. Joseph J. Moldenhauer (Princeton: Princeton University Press, 1972), p. 119.
63. "Walking," in Bode, *Portable Thoreau,* p. 609.
64. *The Maine Woods,* pp. 69–71, 78, 155–56.
65. *Walden,* in Bode, *Portable Thoreau,* pp. 394–400.
66. On the boredom of the tame see, e.g., Irving Howe, "Notes on Mass Culture," in *Mass Culture: The Popular Arts in America,* ed. Bernard Rosenberg and David Manning White (Glencoe, Ill.: Free Press, 1957), pp. 496, 499. See also Eric Larrabee and Rolf Meyersohn, eds., *Mass Leisure* (Glencoe, Ill.: Free Press, 1958).
67. Bode, *Portable Thoreau,* pp. 618–19.
68. "There is something servile in the habit of seeking after a law which we may obey." "Walking," in Bode, *Portable Thoreau,* p. 623.

"The spirit of the American freeman is already suspected to be timid, imitative, tame." "The American Scholar," in Ralph Waldo Emerson, *Selected Prose and Poetry,* 2d ed., ed. Reginald L. Cook (New York: Holt, Rinehart & Winston, 1969), p. 55. See J. S. Mill, *On Liberty* (New York: Penguin Books, Pelican Edition, 1974), pp. 134–35. The ultimate political consequence for societies whose citizens cease to think for themselves is imagined in Yevgeny Zamyatin's antitotalitarian novel, *We* (New York: Bantam Books, 1972), p. 23.

69. A. H. Maslow, *Motivation and Personality* (New York: Harper & Bros., 1954), p. 214. See note 43 above.
70. (New York: Random House, Vintage Books, 1971). See also Michael Murphy, *Golf in the Kingdom* (New York: Dell, Delta Books, 1972).
71. *Zen in the Art of Archery,* p. 74.
72. Ibid., p. 72.

Chapter 4

1. *The Sane Society* (New York: Rinehart & Co., 1955), p. 136; see also Roger Caillois, *Man, Play and Games* (New York: Free Press, 1961), pp. 114–15, 120–22; R. M. MacIver, *The Pursuit of Happiness* (New York: Simon & Schuster, 1955), chap. 6; George A. Pettitt, *Prisoners of Culture* (New York: Charles Scribner's Sons, 1970).
2. Irving Howe, "Notes on Mass Cultures," in *Mass Culture: The Popular Arts in America,* ed. Bernard Rosenberg and David Manning White (Glencoe, Ill.: Free Press, 1957), pp. 496, 499. See also Eric Larrabee and Rolf Meyersohn, eds., *Mass Leisure* (Glencoe, Ill.: Free Press, 1958).
3. John C. Hendee, George H. Stankey, and Robert C. Lucas, *Wilderness Management,* U.S. Forest Service Miscellaneous Publication no. 1365 (October, 1978), pp. 306–7. See also William R. Burch, Jr., "Recreation Preferences As Culturally Determined Phenomena," in *Elements of Outdoor Recreation Planning,* ed. B. L. Driver (Ann Arbor: University of Michigan Press, 1974), pp. 61–87.
4. (New York: Ballantine Books, 1968), p. 62.
5. (New York: Charles Scribner's Sons, 1973), p. 141.
6. The idea goes beyond the usual concept of option value employed by economists. See B. A. Weisbrod, "Collective-Consumption

Services of Individual Consumption Goods," *Quarterly Journal of Economics* 78, no. 3 (August, 1964):471–77. An option value requires us to identify some good or service we do not presently use, but want to retain the opportunity to use. To identify an option value, we must presently recognize that we may want that good or service in the future. Here the concern is with the anterior question, what are the sorts of things to which we ought to attach option values? The failure of ordinary market behavior to reveal the intensity of our wants is explored by the economist Tibor Scitovsky in *The Joyless Economy: An Inquiry into Human Satisfaction and Consumer Dissatisfaction* (New York: Oxford University Press, 1977).

The conventional economic view is that public support is justified if benefits come to us without having to pay for them and if we value those benefits. Baumol and Bowen, "On the Rationale of Public Support," in *Performing Arts—The Economic Dilemma* (New York: Twentieth Century Fund, 1967), p. 369. The quite distinct question the preservationist poses is which benefits provided by parks we should *want* to value.

7. I was introduced to this term by Professor Guido Calabresi of Yale Law School.
8. The content and range of contemporary controversy is illustrated by congressional hearings on National Arts and Humanities Endowment legislation. E.g., U.S., Congress, Senate, *Arts, Humanities and Cultural Affairs Act of 1975: Joint Hearing before the Subcommittee on Select Education, Committee on Education and Labor, House of Representatives, and Special Subcommittee on Arts and Humanities, Committee on Labor and Public Welfare,* 94th Cong., 1st sess., on H.R. 7216 and S. 1800, November 12–14, 1975.

To the extent that publicly supported museums have been persuaded to seek popularity by dramatic exhibitions of celebrated and well-publicized works, there has been an intense critical response within the profession: "The need has never been greater for rigorous standards and their most scrupulous observance," in Brian O'Doherty, ed., *Museums in Crisis* (New York: George Braziller, 1972), p. 83. Hilton Kramer, "The Considerable But Troubling Achievements of Mr. Hoving," *New York Times,* November 11, 1976, sec. 2, p. 1. See also Nathanial Burt, *Palaces for the People* (Boston: Little, Brown & Co., 1977), p. 5. Karl E. Meyer, *The Art of Museum: Power, Money, Ethics* (New York: William Morrow and Co., 1979).

Such controversies are as old as public cultural institutions themselves. E.g., Frederick Rudolph, *The American College and University: A History* (New York: Alfred A. Knopf, 1962), pp. 279–83; Laurence R. Veysey, *The Emergence of the American University* (Chicago: University of Chicago Press, 1965), pp. 100–110; Daniel M. Fox, *Engine of Culture: Philanthropy and Art Museums* (Madison: State Historical Society of Wisconsin, 1963); Sidney Ditzion, *Arsenals of a Democratic Culture* (Chicago: American Library Association, 1947).

9. To some degree we must yield autonomy even in the routine functions of government. Walter Lippman, *The Phantom Public* (New York: Harcourt, Brace & Co., 1925).

10. The typical professional position is summed up in L. Burress, "How Censorship Affects the Schools," Wisconsin Council of Teachers of English Special Bulletin no. 8 (October, 1963): "Censorship is the use of non-professional standards for accepting or rejecting a book. Professional standards are based on the traditional body of literature in English, and assume a familiarity with that literature. Along with the literature is a tradition of literary criticism explaining and evaluating the literature. Though a group of professionally trained people may well disagree on the merits of a given book, a working consensus can be obtained, subject to continued debate in the forum provided by literary journals. That is . . . a public process, and can be joined by any interested person who will familiarize himself with the rules of the game."

    Regarding professional group values as a powerful constraint and influence on personal values, see, e.g., Daniel Katz and Robert L. Kahn, *The Social Psychology of Organizations* (New York: John Wiley & Sons, 1966), pp. 51–57; Jerome Bruner et al., "Personal Values as Selective Factors in Perception," *Journal of Abnormal and Social Psychology* 43 (1948):142, 154; Mortimer B. Smith et al., *Opinions and Personality* (New York: John Wiley & Sons, 1956), pp. 41–43; H. C. Kelman, "Three Processes of Social Influence," in *Attitudes*, ed. Marie Johoda and Neil Warren (Baltimore, Penguin Books, 1966), p. 153.

11. The appropriate legal ground for determining the scope of a teacher's or librarian's academic freedom has been a subject of intense controversy among legal scholars, especially at the preuniversity level, where formal authority over curriculum is vested in a school board. Where the teacher is plainly in the mainstream of professional and critical judgment, rather than acting idiosyncratically or officiously testing the bounds of community decency, the

inclination to follow the intellectual tradition in the profession is very strong, whatever the formal ground for the decision. E.g., Keefe v. Geanakos, 418 F.2d 359 (1st Cir. 1969); Parducci v. Rutland, 316 F. Supp. 352 (M.D. Ala. 1970); Mailloux v. Kiley, 436 F.2d 565 (1st Cir. 1971), 448 F.2d 1242 (1st Cir. 1971); Minarcini v. Strongsville City School Dist., 541 F.2d 577 (6th Cir. 1976); Epperson v. Arkansas, 393 U.S. 97 (1968). Compare Presidents Council v. Community School Board, 457 F.2d 289 (2d Cir. 1971), *cert. denied,* 409 U.S. 998 (1972), criticized in O'Neil, "Libraries, Liberties and the First Amendment," *University of Cincinnatti Law Review* 42(1973):209. For a reference to the extensive legal literature on the subject see Goldstein, "The Asserted Right of Public School Teachers to Determine What They Teach," *University of Pennsylvania Law Review* 124(1976):1293.

12. (San Francisco: Jossey-Bass, 1975).
13. Ibid., p. 196.
14. *On Becoming a Person* (New York: Houghton Mifflin Co., 1961). A more detailed explanation of Rogers's theory appears in "A Theory of Therapy, Personality, and Interpersonal Relationships, as Developed in the Client-Centered Framework," in *Psychology: A Study of a Science,* ed. Sigmund Koch, vol. 3 (New York: McGraw Hill, 1959). Rogers's views and their place in modern psychology are described in Calvin S. Hall and Gardner Lindzey, *Theories of Personality,* 2d ed. (New York: John Wiley & Sons, 1970).
15. *Motivation and Personality* (New York: Harper & Bros., 1954), p. 214. See also *The Farther Reaches of Human Nature* (New York: Viking, 1971).
16. (Cambridge, Mass.: Harvard University Press, 1971).
17. Ibid., pp. 426–27. A view rather like Rawls's appears in the economist Tibor Scitovsky's *The Joyless Economy: An Inquiry into Human Satisfaction and Consumer Dissatisfaction* (New York: Oxford University Press, 1977). For an example of current ways of studying user satisfaction see, e.g., U.S., Department of Agriculture, Forest Service, *Proceedings: River Recreation Management and Research Symposium,* January 24–27, 1977, Minneapolis, Minn., North Central Forest Experiment Station, General Technical Report NC-28, pp. 359–64.
18. E.g., Reuben Fine, *The Psychology of the Chess Player* (New York: Dover Publications, 1967), noting Ernest Jones's classic paper, "The Problem of Paul Morphy," read to the British Psychoanalytical Society in 1930. Jones's paper provoked a number of psycholog-

ical studies of chess players, all of which concur in describing "hostile impulses," "sense of overwhelming mastery," "love of pugnacity," and "competitiveness."

19. There seems to be considerable agreement that winning a game, simply to prevail, diminishes the depth of the player's satisfaction. But whether the structure of the game itself is significant remains an unsettled, and largely unexplored, question. See Johan Huizinga, *Homo Ludens: A Study of the Play Element in Culture* (London: Routledge & Kegan Paul, 1949), p. 197, criticizing the systemization and professionalization of sports. See also Allen Guttman, *From Ritual to Record: The Nature of Modern Sports* (New York: Columbia University Press, 1978), p. 9, discussing the thesis that games are structured devices for exhibiting mastery, but making no distinction between zero-sum games and others which permit but do not depend on winning against an opponent.

Chapter 5

1. U.S., Department of the Interior, National Park Service, *Guadalupe Mountains National Park, Environmental Assessment, Development Concept Plan* (September, 1975).
2. U.S., Department of the Interior, National Park Service, *Guadalupe Mountains National Park, Master Plan* (March 29, 1973).
3. E.g., U.S., Department of the Interior, National Park Service, Glen Canyon National Recreation Area, *Proposed General Management Plan, Wilderness Recommendation, Road Study Alternatives, Final Environmental Statement* (July, 1979), p. 29: ". . . yearly visitation from 1962, when the recreation area was established, through 1977. Linear trend analysis for the 16-year period yields an average annual increase of about 25 percent per year. . . . "
4. The discussion of the Olympic National Park controversy is taken from the following sources: U.S., Department of the Interior, National Park Service, "National Park Service War Work, December 7, 1941 to June 30, 1944, and June 30, 1944 to October 1, 1945," mimeographed; memorandum, Newton B. Drury, director, National Park Service, to secretary of the interior, December 18, 1942; Abe Fortas, acting secretary of the interior, to Senator Homer T. Bone, December 17, 1942; confidential memorandum, superintendent, Olympic National Park, to director, National Park Service, June 11, 1941; supplemental memorandum on supply of Sitka spruce, from W. T. Andrews, consultant, November 17, 1940; John Ise, *Our National Park Policy: A Critical History*

(Baltimore: Johns Hopkins University Press, 1961), pp. 392–93; annual reports of the secretary of the interior for 1942, 1943, 1944; Robert Shankland, *Steve Mather of the National Parks* (New York: Alfred A. Knopf, 1951), p. 306; Elmo R. Richardson, *Dams, Parks and Politics* (Lexington: University of Kentucky Press, 1973), pp. 10–13; "Will the Needs of War Require Loss of Olympic National Park," *The Living Wilderness,* May, 1943, pp. 26–27; Edgar B. Nixon, ed., *Franklin D. Roosevelt and Conservation 1911–1945,* 2 vols. (Hyde Park, N.Y.: General Services Administration, 1957), 2:559–60, 572–73, 578–79.

5. Developments on national forest land sometimes affect the parks quite directly. The Jackson Hole ski development in Wyoming has generated urbanizing pressures that are now being felt on land directly adjacent to Grand Teton National Park. The description of the Mineral King controversy is taken from these sources: Jeanne Ora Nienaber, "Mineral King: Ideological Battleground for Land Use Disputes" (Ph.D. diss., University of California, Berkeley, n.d.); Susan R. Schrepfer, "Perspectives on Conservation: Sierra Club Strategies in Mineral King," *Journal of Forest History* 20, no. 5 (October, 1976):176–90; Commentary, "Mineral King Goes Downhill," *Ecology Law Quarterly* 5 (1976):555–74; Peter Browning, "Mickey Mouse in the Mountains," *Harper's* 244 (March, 1972):65, 245 (August, 1972):102; U.S., Department of Agriculture, Forest Service Region 5, *Mineral King Final Environmental Statement* (San Francisco, February 26, 1976).
6. *Sierra Club Bulletin* 52, no. 11 (November, 1967):7. Sierra Club v. Morton, 405 U.S. 727 (1972).
7. The Mineral King area has now been added to the adjacent Sequoia National Park, with a mandate that will prevent its development as a ski resort. Public Law 95–625, § 314, 92 Stat. 3479 (November 10, 1978).
8. *Los Angeles Times,* January 13, 1978, p. 36.
9. Schrepfer, "Perspectives on Conservation," p. 184.
10. Calculating demand, as an economic matter, is complex and difficult. Marion Clawson and Jack L. Knetch, *Economics of Outdoor Recreation* (Baltimore: Johns Hopkins University Press, 1966); U.S., Department of the Interior, Bureau of Outdoor Recreation, *Outdoor Recreation: A Legacy for America,* appendix A, "An Economic Analysis" (December, 1973). Economists are concerned with observed willingness to pay as a measure of benefits from the project. "Mineral King Valley: Demand Theory and Resource Valuation,"

in John V. Krutilla and Anthony C. Fisher, *The Economics of Natural Environments* (Baltimore: Johns Hopkins University Press, 1975), chap. 8, pp. 189–218. My concern here is with a complication of that measure from the perspective of a supplier who is willing to forego some benefits, measured by consumer willingness to pay, for others measured by collectively articulated goals; and who thus feels obliged to meet some, but not all, conventional demand. Economic analysts may ignore such distinctions: "Suppose (winter) visitors to a ski site are going for other purposes as well, simply to play in the snow, say, or to enjoy the social environment. Is this a problem? We don't think so." Krutilla and Fisher, *Economics of Natural Environments,* p. 198.

The existence of a given recreational demand does not itself demonstrate that national park policy should be committed to its fulfillment, any more than a demand for cosmetic surgery has to be fulfilled by a public medical care policy. Likewise, we might well decide upon a public policy of building hospitals, and not hotels, though there is a considerable demand for hotel rooms, and vacationers might be willing and able to outbid sick people for available beds. Collectively, as owners of the public lands and citizens, we can have a collective desire different from the sum of market demands made by each of us individually. We may decide to forego the greatest dollar return on our property in favor of some use that we believe provides a greater return in satisfaction.

11. See U.S., Department of Agriculture, Forest Service, *Environmental Statement (Final), Mount Hebgen, Management Alternatives, Gallatin National Forest,* USDA-FS-R1(11) FES-Adm-76-25 (May 13, 1977); petition of Montana Wilderness Association, Montana Wildlife Federation, the Environmental Information Center, Wilderness Society and the Sierra Club (United States, Department of Agriculture, Before the Regional Forester, Region 1, 1977); Letter of Final Administrative Determination, John R. McGuire, chief, Forest Service, to James H. Goetz (August 1, 1978).
12. Bruce L. Nurse to Lewis E. Hawkes, forest supervisor, Gallatin National Forest, February 21, 1977, in the petition of Montana Wilderness Association (see note 11 above), exhibit 10.
13. Morton Lund, "The Sage of Sun Valley," *Ski Area Management* Spring, 1972, p. 35.
14. Clay R. Simon to Lewis E. Hawkes, forest supervisor, Gallatin National Forest, May 27, 1975, in the petition of Montana Wilderness Association (see note 11 above), exhibit 9.

15. See note 12, above.
16. *Environmental Statement,* p. B-15.
17. See note 12 above. Charles J. Cicchetti, Joseph J. Seneca, and Paul Davidson, *The Demand and Supply of Outdoor Recreation* (New Brunswick, N.J.: Rutgers University Press, 1969); *Environmental Statement,* p. B-19.
18. Such facilities require "capability of handling large crowds efficiently (volume is everything in such an operation) [and] keeping [the visitor] in a happy (i.e., spending) frame of mind." Richard Schickel, *The Disney Version* (New York: Avon Books, 1968), p. 114.
19. William E. Shands and Robert G. Healy, *The Lands Nobody Wanted* (Washington, D.C.: Conservation Foundation, 1977), pp. 47–48.
20. U.S., Congress, House, *National Park Service Planning and Concession Operations: Joint Hearing Before Certain Subcommittees of the Committee on Government Operations and the Permanent Select Committee on Small Business,* 93d Cong., 2d sess., December 20, 1974.
21. U.S., Department of the Interior, National Park Service, *Yosemite National Park Preliminary Draft Master Plan* (August 12, 1974).
22. U.S., Congress, House, *National Park Service Planning and Concession Operations,* pp. 226–27, 235–36, 238–41, 290–95.
23. The most recent Park Service plan for Yosemite recommends some reduction in the existing facilities at the park. U.S., Department of the Interior, National Park Service, *Final General Management Plan for Yosemite National Park* (January, 1980). The draft plan of August, 1978, was more far-reaching.
24. The Park Service itself at times promotes development in order to generate demand for visits at sparsely used parks, focusing on what it take to attract casual tourists passing by. In the now-shelved master plan for Guadalupe Mountains National Park in Texas (recommended March 29, 1973), park service planners observed: "Visitor use . . . will be seriously impeded until motels, restaurants and campgrounds become available within a convenient distance [p. 38]." To make the park "a magnet for visitors [p. 2]," a tramway to the top of Guadalupe Peak was recommended as an "educational and inspirational experience [p. 37]." See notes 1 and 2 above.
25. Alta's economic viability may depend on its proximity to day users from nearby Salt Lake City. If skiing without resort facilities is not profitable in most circumstances, the question for public policy

would be whether we want to subsidize that activity, as we do many others.

26. Steven V. Roberts, "Visitors Are Swamping the National Parks," *New York Times,* September 1, 1969, p. 15.
27. *The Revolt of the Masses* (New York: W. W. Norton & Co., 1957), p. 58.
28. Ibid., pp. 57–59: "The world which surrounds the new man . . . incites his appetite, which in principle can increase indefinitely . . . two fundamental traits: the free expansion of his vital desires . . . and his radical ingratitude towards all that has made possible the ease of his existence. These traits together make up the well-known psychology of the spoilt child. . . . To spoil means to put no limit on caprice, to give one the impression that everything is permitted to him and that he has no obligations."

Chapter 6

1. Act of October 21, 1976, 90 Stat. 2733, Public Law 94–578, Title III, § 301, repealing a provision directing the construction of a road from the Chincoteague–Assateague Island Bridge to an area in the wildlife refuge, for recreation purposes.
2. U.S., Department of the Interior, National Park Service, *Gateway National Recreation Area, General Management Plan, Discussion Draft* (September, 1976), pp. 14, 34. A decision paper was issued in April, 1978, and a general management plan and final environmental statement in 1979.
3. Ibid., p. 64.
4. One of the most encouraging Park Service expressions of intent is the "interpretive concepts" section of the *Draft Master Plan for Acadia National Park* (May, 1976), pp. 22–26.
5. In his journal for October 23, 1837, Ralph Waldo Emerson said: "Culture is not the trimming and turfing of gardens, but the showing [of] the true harmony of the unshorn landscape . . ." E. W. Emerson and W. E. Forbes, eds., *Journals of Ralph Waldo Emerson,* vol. 4 (Boston: Houghton Mifflin Co., 1910) p. 340.
6. The significance of Thoreau's contrasting of tameness and wildness may be seen even in these seemingly minor matters. In *The Disney Version* (New York: Avon Books, 1968), p. 39, Richard Schickel finds in the Disney films "a drive . . . toward . . . multiple reductionism; wild things and wild behavior were often made comprehensible by converting them into cuteness, mystery was explained by a

joke and . . . terror was resolved by . . . a discreet averting the camera's eye from the natural processes. . . . [T]here is something deeper than [money] at work in this national passion to tame. . . . "

7. Jeremy Bernstein, *Mountain Passages* (Lincoln: University of Nebraska Press, 1978), p. 66. See also, Organization for Economic Co-operation and Development, Environment Directorate, "The Growth of Ski-Tourism and Environmental Stress in Switzerland," ENV/TOUR/78.5, Paris, April 25, 1978.

Chapter 7

1. Act of September 3, 1964, Public Law 88–577, 78 Stat. 890, 16 U.S.C. § 1131.
2. Francoise Leydet, *The Grand Canyon* (San Francisco: Sierra Club, 1964). See also T. H. Watkins et al., *The Grand Colorado: The Story of a River and Its Canyons* (Palo Alto: American West Publishing Co., 1969), p. 270.
3. Eliot Porter, *The Place No One Knew: Glen Canyon on the Colorado* (San Francisco: Sierra Club, 1963); see National Parks Ass'n v. Udall, Civil No. 3904–62 (U.S. Dist. Ct. 1962). The earlier, precedent-setting battle over the proposed Dinosaur Monument–Echo Park Dam is described in John Ise, *Our National Park Policy: A Critical History* (Baltimore: Johns Hopkins University Press, 1961), pp. 476–80.
4. J. W. Powell, *The Exploration of the Colorado River and Its Canyons* (New York: Dover Publications, 1961), a republication of the original work published in 1895 under the title *Canyons of the Colorado.* For a fine biography of Powell, see Wallace Stegner, *Beyond the Hundredth Meridian: John Wesley Powell and the Second Opening of the West* (Boston: Houghton Mifflin Co., 1954).
5. U.S., Department of the Interior, National Park Service, *Colorado River Management Plan, Grand Canyon National Park, Arizona* (December 20, 1979), p. 1. See idem, "*Final Environmental Statement, Proposed Colorado River Management Plan, Grand Canyon National Park, Arizona* (n.d.).
6. *Colorado River Management Plan,* pp. 8–9.
7. *Final Environmental Statement,* p. II–41. Detailed information about river controversies appears in U.S., Department of Agriculture, Forest Service, *Proceedings: River Recreation Management and Research Symposium,* January 24–27, 1977, Minneapolis, Minn., North Central Forest Experiment Station, General Technical Report NC-28).

8. E.g., Fred B. Eiseman, Jr., "Who Runs the Grand Canyon?", *Natural History* 87, no. 3 (March, 1978):83–93; Liz Hymans, "The Flow of Wilderness," *Not Man Apart* (Friends of the Earth) 8, no. 5 (Mid-March, 1978).
9. Wilderness Public Rights Fund v. Kleppe, Eisemann v. Kleppe, 13 ERC 2094 (9th Cir. 1979); Western River Expeditions, Inc. v. Morton, Civ. No. C-125-B (U.S. Dist. Ct. Utah), Order of June 4, 1973; Grand Canyon National Park v. Stitt, No. 77–722 (U.S. Dist. Ct. Ariz.) (filed September 19, 1977, pending).
10. See note 5 above.
11. It is appropriate for the Park Service to put some maximum limit on the duration of visits. There is an administrable line between leisure and monopoly.
12. Colorado Outward Bound School River Trips, for example, have a very different perspective. See their undated mimeo pamphlet, "Which Trip for You." They do not run trips in Grand Canyon National Park.
13. E.g., "Jack Currey's Western River Expeditions" (1977 catalog), p. 28; American River Touring Association, "River Adventure: 1978," p. 21.
14. Adventure Bound, Inc., "River Expeditions 1977" (unpaged). This concessioner does not run trips in Grand Canyon National Park; its area is from Dinosaur National Monument and from above Arches National Park down to Glen Canyon National Recreation Area.
15. "Jack Currey's," p. 28.
16. Ibid.
17. Ibid, p. 4.
18. American River Touring 1979 catalog, p. 9 (7 days, Lee's Ferry to Diamond Creek).
19. See Roggenbuck and Schreyer, "Relations Between River Trip Motives and Perception of Crowding, Management Preference and Experience Satisfaction," in U.S., Department of Agriculture, *Proceedings: River Recreation Management and Research Symposium,* pp. 359–64. Mordechai Shechter and Robert C. Lucas, *Simulation of Recreational Use for Park and Wilderness Management* (Baltimore: Johns Hopkins University Press, 1978), chap. 9.
20. When such conflicts arise, policy considerations other than recreation may be determinative. For example, a decision might be made to retain an area as wilderness for its resource values, or to retain scientific or archeological features, even though to do so would

prevent meeting recreational demands that would have prevailed if the only conflict were between competing recreational policies. So it is quite possible that—for reasons unrelated to recreation policy—a place like Grand Canyon would be closed to high density service recreation even though there is a shortage of service recreation areas and an "over-abundance" of pristine areas. Similarly, a determination of strong industrial need could justify removing an area from management for reflective recreation, and permit service recreation, though there are already abundant opportunities for service recreation. In such a case, recreation policy generally would be subordinated to some other priority (such as the need to mine a greatly needed mineral). Nothing in this book is meant to suggest that recreation policy, per se, must prevail over other policies. It is directed solely to conflicts among competing recreational policies.

21. E.g., Smoky Mountain Field School, a series of extended summer workshops cosponsored by the National Park Service and the University of Tennessee Division of Continuing Education. Cf. the Yellowstone Institute, run by the Yellowstone Library and Museum Association, and the Yosemite Institute's School Weeks program.
22. *Woodall's Trailer and RV Travel,* January, 1978, p. 5.
23. See note 4 above.
24. Richard Jones's Worldwide River Expeditions, "Rivers USA 1977," p. 3.
25. "1977 Colorado River & Trail Expeditions, Inc.," p. 1.
26. American River Touring, p. 21.
27. A nice description of going it alone in the canyon today is "Down the River with Major Powell," in Edward Abbey, *The Journey Home* (New York: E. P. Dutton, 1977), chap. 17.
28. Title 2300, Recreation Management, ¶ 2331.11c, exhibit I.
29. The Forest Service had an analogous policy of refraining from cutting commercial timber right up to highways so that citizens—offended by the idea of murdering trees, though they used wood products—would be screened from unwanted reality. Harold K. Steen, *The U.S. Forest Service: A History* (Seattle: University of Washington Press, 1976), p. 159.
30. The truly self-defining individual can pay a high price in loneliness for his or her inner freedom, and that price is nowhere more poignantly portrayed than in Thoreau's journals. See Odell Shepard, *The Heart of Thoreau's Journals* (New York: Dover Publications,

1961), pp. 142, 172–73, 175. "If an individual gives up his distinctiveness in a group . . . he does it because he feels the need of being in harmony with them rather than in opposition to them." Sigmund Freud, *Group Psychology and the Analysis of the Ego,* ed. and trans. James Strachey (New York: W. W. Norton & Co., 1959), p. 24.

31. "The master had but to look at him, when this young man would fling himself back as though struck by lightening, place his hands rigidly at his sides, and fall into a state of military somnambulism, in which it was plain to any eye that he was open to the most absurd suggestion that might be made to him. He seemed quite content in his abject state, quite pleased to be relieved of the burden of voluntary choice." Thomas Mann, "Mario and the Magician," in *Stories of Three Decades* (New York: Alfred A. Knopf, 1936), p. 560.

"I tell Thee that man is tormented by no greater anxiety than to find someone quickly to whom he can hand over that gift of freedom with which the ill-fated creature is born. But only one who can appease their conscience can take over their freedom." Fyodor Dostoevski, *The Brothers Karamazov* (New York: Modern Library, 1950), p. 302.

Chapter 8

1. E.g., Chez des Amis, 139 W. 87th St., New York, New York.
2. René Jules Dubos, "The Genius of the Place," Tenth Annual Horace M. Albright Conservation Lectureship, University of California at Berkeley, School of Forestry and Conservation, February 26, 1970.
3. Clara Barrus, *Our Friend John Burroughs* (Boston: Houghton Mifflin Co., 1914), p. 131.

Bibliographic Notes

The standard, and excellent, history of the national parks is John Ise, *Our National Park Policy: A Critical History* (Baltimore: Johns Hopkins University Press, 1961). A history of the national park idea, is Alfred Runte, *National Parks: The American Experience* (Lincoln: University of Nebraska Press, 1979). For a view of the parks by an insider during the early days of the National Park Service, see Robert Sterling Yard, *The Book of the National Parks* (New York: Charles Scribner's Sons, 1919). A fine descriptive book is Freeman Tilden, *The National Parks* (New York: Alfred A. Knopf, 1968).

For more detailed information about park history consult Paul Buck, *The Evolution of the National Park System of the United States* (Washington, D.C.: Government Printing Office, 1946); Donald C. Swain, "The Passage of the National Park Service Act of 1916," *Wisconsin, Magazine of History* 50 (1966):4; Alfred Runte, "The National Park Idea: Origins and Paradox of American Experience," *Journal of Forest History* 21, no. 2 (April, 1977):64; Joseph L. Sax, "America's National Parks: Their Principles, Purposes and Prospects," *Natural History* 85, no. 8 (October, 1976):57.

The history of recreation policy in the national forests is best detailed in the Ph.D. dissertation of James P. Gilligan, "The Development of Policy and Administration of Forest Service Primitive and Wilderness Areas in the Western United States" (University of Michigan, 1953); see also Harold K. Steen, *The U.S. Forest Service: A History* (Seattle: University of Washington Press, 1976); Donald F. Cate, "Recreation and the U.S. Forest Service: Organizational Response to Changing Demands" (Ph.D. diss., Stanford University, 1963). The leading scholarly book on national forests is Samuel T. Dana, *Forest and Range Policy*

(New York: McGraw-Hill, 1956). There is much less literature dealing with recreation on Bureau of Land Management areas. Marion Clawson, *The Bureau of Land Management* (New York: Praeger, 1971); *Preserving Our Natural Heritage,* vol. 1, *Federal Activities,* prepared for the United States Department of the Interior, National Park Service, by the Nature Conservancy (Washington, D.C.: Government Printing Office, 1977).

There are several fine books on the history of American attitudes toward nature that sharply illuminate the ideas underlying the establishment of national parks. The best of these are Hans Huth, *Nature and the American: Three Centuries of Changing Attitudes* (Lincoln: University of Nebraska Press, 1972), with an excellent chapter on the founding of Yosemite National Park; Roderick Nash, *Wilderness and the American Mind* (New Haven: Yale University Press, 1973); Leo Marx, *The Machine in the Garden: Technology and the Pastoral Ideal in America* (New York: Oxford University Press, 1964); Cecilia Tichi, *New World, New Earth: Environmental Reform in American Literature from the Puritans Through Whitman* (New Haven: Yale University Press, 1979). See also Norman Foerster, *Nature in American Literature* (New York: Russell & Russell, 1923); Roderick Nash, ed., *The American Environment: Readings in the History of Conservation* (Reading, Mass.: Addison-Wesley, 1968).

The symbolism of the untamed West, parallel to but distinct from movements for nature preservation, is superbly limned in Henry Nash Smith, *Virgin Land: The American West as Symbol and Myth* (Cambridge, Mass.: Harvard University Press, 1950). See also Bernard DeVoto, *Mark Twain's America* (Boston: Houghton Mifflin Co., Sentry, 1967). A European perspective is provided by Raymond Williams, *The Country and the City* (New York: Oxford University Press, 1973).

The wilderness movement is a distinct subspecies of the larger American involvement with nature. Among useful books devoted solely to it are Michael Frome, *Battle for the Wilderness* (New York: Praeger, 1974), and Donald N. Baldwin, *The Quiet Revolution: The Grass Roots of Today's Wilderness Preservation Movement* (Boulder, Colo.: Pruett Publishing Co., 1972). A wide range of references are provided in John C. Hendee, George H. Stankey, and Robert C. Lucas, *Wilderness Management,* U.S., Department of Agriculture, Forest Service, Miscellaneous Publication no. 1365 (Washington, D.C.: Government Printing Office, 1978).

The early histories of national parks themselves are not generally highly illuminating. Yosemite and Yellowstone are the exceptional cases. On Yosemite, see Hans Huth, "Yosemite: The Story of an Idea,"

Sierre Club Bulletin 33, no. 3 (1948):47; Shirley Sargent, *Galen Clark, Yosemite Guardian* (San Francisco: Sierra Club, 1964); Carl Parcher Russell, *One Hundred Years in Yosemite* (Berkeley and Los Angeles: University of California Press, 1947); Holway R. Jones, *John Muir and the Sierra Club: The Battle for Yosemite* (San Francisco: Sierra Club, 1965). On Yellowstone, see Aubrey L. Haines, *The Yellowstone Story: A History of Our First National Park,* 2 vols. (Yellowstone Library and Museum Association in cooperation with Colorado Associated University Press, 1977); Aubrey L. Haines, *Yellowstone National Park: Its Exploration and Establishment* (Washington, D.C.: Government Printing Office, 1974); H. Duane Hampton, *How the U.S. Cavalry Saved Our National Parks* (Bloomington: Indiana University Press, 1971).

Hans Huth, *Nature and the American,* has a chapter on Yellowstone, Yosemite, and Grand Canyon. Frank Graham, *The Adirondack Park* (New York: Alfred A. Knopf, 1977), is an excellent history of a state park. Carolyn de Vries, *Grand and Ancient Forest: The Story of Andrew P. Hill and Big Basin Redwood State Park* (Fresno, Calif.: Valley Publishers, 1978) also has useful information. Typical of other books on the national parks are Carlos C. Campbell, *Birth of a National Park in the Great Smoky Mountains,* rev. ed. (Knoxville: University of Tennessee Press, 1969) and Robert Treuer, *Voyageur Country: A Park in the Wilderness* (Minneapolis: University of Minnesota Press, 1979). See also Peter Wild, *Enos Mills* (Rocky Mountain National Park) (Boise, Idaho: Boise State University, 1979).

Early visitor attitudes and expectations are revealed in Earl Pomeroy, *In Search of the Golden West: The Tourist in Western America* (New York: Alfred A. Knopf, 1957.)

There are many individuals who gave substance to a park idea through their work or their writing. A sketch of some of the major figures is found in Peter Wild, *Pioneer Conservationists of Western America* (Missoula, Mont.: Mountain Publishing Co., 1979).

John Muir, more than any other individual, symbolizes the parks in the American mind. There is no first-rate Muir biography, unfortunately. The best available work is Linnie Marsh Wolfe, *Son of the Wilderness: The Life of John Muir* (New York: Alfred A. Knopf, 1945). Muir's partial autobiography, *The Story of My Boyhood and Youth* (Madison: University of Wisconsin Press, 1965), is fascinating. Holway R. Jones, *John Muir and the Sierra Club,* is also very informative, though its scope is limited. There are several good dissertations on Muir: Edith Jane Hadley, *John Muir's Views of Nature and Their Consequences* (Ph.D.

diss., University of Wisconsin, 1956) and Daniel Barr Weber, *John Muir: The Function of Wilderness in an Industrial Society* (Ph.D. diss., University of Minnesota, 1964). Though Muir's writings are widely scattered in the periodical literature, there is a fine bibliography: William F. Kimes and Maymie B. Kimes, *John Muir: A Reading Bibliography* (Palo Alto: William P. Wreden, 1977). The most accessible selection of Muir's articles is found in Edwin Way Teale, ed., *The Wilderness World of John Muir* (Boston: Houghton Mifflin Co., 1952).

John Burroughs, a contemporary of Muir's, was an extremely popular nature writer, and was in a sense a domestic counterpart to Muir, the explorer and mountain man. For a sense of Burroughs, see Clifton Johnson, *John Burroughs Talks* (Boston: Houghton Mifflin Co., 1922); Clara Barrus, *Our Friend John Burroughs* (Boston: Houghton Mifflin Co., 1914).

There are two excellent biographies of Frederick Law Olmsted: Elizabeth Stevenson, *Park Maker: A Life of Frederick Law Olmsted* (New York: Macmillan Co., 1977) and Laura Wood Roper, *FLO: A Biography of Frederick Law Olmsted* (Baltimore: Johns Hopkins University Press, 1973). Almost all Olmsted's important writings on nature park issues are found in rather obscure places. His great article "The Yosemite Valley and the Mariposa Big Trees" appears in *Landscape Architecture* 44, no. 1 (1953):17. His writings on Niagara Falls are in "Notes by Mr. Olmsted," *The Special Report of the New York State Survey on the Preservation of the Scenery of Niagara Falls and Fourth Annual Report on the Triangulation of the State for the Year 1879* (Albany: Charles von Benthuysen & Sons, 1880) p. 27; and "Report of Messrs. Olmsted and Vaux," *Supplemental Report of the Commissioners of the State Reservation at Niagara Transmitted to the Legislature January 31, 1887* (Albany: Argus Co., 1887), p. 21. The fascinating movement to save Niagara is detailed in Alfred Runte, "Beyond the Spectacular: The Niagara Falls Preservation Campaign," *New York State Historical Society Quarterly* 57 (January, 1973):30; "The Movement for the Redemption of Niagara," *New Princeton Review* (New York: A. C. Armstrong & Sons, 1886), 1:233; and Charles M. Dow, *The State Reservation at Niagara: A History* (Albany: J. B. Lyon Co., 1914). See also *Public Parks and the Enlargement of Towns* (Cambridge, Mass.: Riverside Press, 1870; reprint ed., New York: Arno Press and The New York Times, 1970).

There is as yet, unhappily, no full biography of Aldo Leopold, the major figure in the establishment of forest wilderness. See Susan L. Flader, *Thinking Like a Mountain: Aldo Leopold and the Evolution of an*

Ecological Attitude Toward Deer, Wolves and Forests (Lincoln: University of Nebraska Press, 1974). Like John Muir, Aldo Leopold scattered much of his writing in journals. A characteristic article is "The Wilderness and Its Place in Forest Recreational Policy," *Journal of Forestry* 19, no. 7 (November, 1921):718. Leopold's best known book is *A Sand County Almanac With Other Essays on Conservation from Round River* (New York: Oxford University Press, 1966).

The best article of Robert Marshall, a founder of the Wilderness Society, is "The Problem of Wilderness," *Scientific Monthly* 30 (February, 1930):141. He also wrote on his travels in Alaska: *Alaska Wilderness Exploring the Central Brooks Range* (Berkeley and Los Angeles: University of California Press, 1973).

Among contemporary writers, none is more influential or widely read than Edward Abbey. The best of his nonfiction books is *Desert Solitaire* (New York: Random House, Ballantine Books, 1971); see also *The Journey Home* (New York: E. P. Dutton, 1977); *Abbey's Road* (New York: E. P. Dutton, 1979); and with Philip Hyde, *Slickrock: The Canyon Country* (San Francisco: Sierra Club, 1971).

The American nature-writing genre is extensive, and no selection can do more than suggest the range and variety of its influence on preservationist thinking. Sigurd Olson, *Reflections from the North Country* (New York: Alfred A. Knopf, 1976); Joseph Wood Krutch, *The Voice of the Desert* (New York: William Sloane Assoc., 1975); Gary Snyder, *Turtle Island* (New York: New Directions, 1974); John McPhee, *Coming Into the Country* (New York: Bantam Books, 1979); Wallace Stegner, "The War Between the Rough Riders and the Bird Watchers," *Sierra Club Bulletin* 44, no. 4 (May, 1959); and Margaret E. Murie, *Two in the Far North* (Anchorage, Alaska: Northwest Publishing Co., 1978) reveal the variousness of contemporary writing.

For a view of another contemporary leader in the parks and nature preservation movement, David Brower of Friends of the Earth, see John McPhee, *Encounters with the Archdruid* (New York: Farrar, Straus and Giroux, 1971).

Theodore Roosevelt is the most prominent of those who played a central role in the political history of the parks. There is, of course, a superabundance of writing on Roosevelt, but a few books are focused on his role in conservation history. Paul Russell Cutright, *Theodore Roosevelt the Naturalist* (New York: Harper & Bros., 1956) and Francis Cevrier Guittard, *Roosevelt and Conservation* (Ph.D. diss., Stanford University, 1930). An excellent book on the early Roosevelt is Edmund Morris,

The Rise of Theodore Roosevelt (New York: Coward, McCann & Geoghegan, 1979).

Both the first and second directors of the National Park Service have been the subject of useful biographies. Robert Shankland, *Steve Mather of the National Parks* (New York: Alfred A. Knopf, 1951) and Donald C. Swain, *Wilderness Defender: Horace M. Albright and Conservation* (Chicago: University of Chicago Press, 1970). One of the great American figures, though he is not usually associated with parklands or recreation, is John Wesley Powell, the explorer of the Colorado River and head of the United States Geological Survey. His thinking has, though indirectly, been enormously influential. Wallace Stegner's fine biography of Powell is *Beyond the Hundredth Meridian: John Wesley Powell and the Second Opening of the West* (Boston: Houghton Mifflin Co., 1954).

Towering above all others in influence and in the depth of his writing is Thoreau, and standing beside him is Emerson. The best selection of Thoreau's work appears in Carl Bode, ed., *The Portable Thoreau* (New York: Viking, 1975); and in Odell Shepard, ed., *The Heart of Thoreau's Journals* (New York: Dover Publications, 1961). Of course there is no substitute for reading Thoreau *in extenso.* A fine book about him is Walter Harding, *The Days of Henry Thoreau* (New York: Alfred A. Knopf, 1965). A good selection of Emerson's writing appears in Ralph Waldo Emerson, *Selected Prose and Poetry,* 2d ed., ed. Reginald L. Cook (New York: Holt, Rinehart & Winston, 1969).

Many of the ideas about the importance of parks as public institutions found in the writings of conservation movement figures are paralleled in nineteenth-century thinking about places like universities, libraries, and museums. See, for example, Frederick Rudolph, *The American College and University: A History* (New York: Alfred A. Knopf, 1962); Laurence R. Veysey, *The Emergence of the American University* (Chicago: University of Chicago Press, 1965); Daniel M. Fox, *Engines of Culture: Philanthropy and Art Museums* (Madison: State Historical Society of Wisconsin, 1963); Sidney Ditzion, *Arsenals of a Democratic Culture* (Chicago: American Library Association, 1947); Nathanial Burt, *Palaces for the People,* (Boston: Little, Brown & Co., 1977); William B. Ashley, "The Promotion of Museums," *Proceedings of the American Association of Museums* 7 (1913):39.

Books about the recreational experience are legion. On fishing, the great book is Izaak Walton and Charles Cotton, *The Compleat Angler* (London: Navarre Society, 1925). An ample bibliography appears in Arnold Gingrich, *The Well-Tempered Angler* (New York: Alfred A.

Knopf, 1966). A beautiful American fictional story on fishing is Ernest Hemingway's "Big Two-Hearted River," in *In Our Time* (New York: Charles Scribner's Sons, 1970). I also find Norman MacLean's "A River Runs Through It" splendidly evocative, *A River Runs Through It and Other Stories* (Chicago: University of Chicago Press, 1976).

The most stimulating book on hunting is José Ortega y Gasset's *Meditations on Hunting* (New York: Charles Scribner's Sons, 1972), a book that is best understood when read in conjunction with the same author's *The Revolt of the Masses* (New York: W. W. Norton & Co., 1957). The great American hunting story is William Faulkner's "The Bear" in *The Portable Faulkner,* ed. Malcolm Cowley (New York: Viking Press, 1967). See also Paul Shepard, *The Tender Carnivore and the Sacred Game* (New York: Charles Scribner's Sons, 1973); John G. Mitchell, "Bitter Harvest," *Audubon* 81, nos. 3, 4, 5, 6 (1979):50, 64, 88, 104.

There is no single mountain climbing book that stands alone, as does Walton and Cotton on fishing. Maurice Herzog, *Annapurna* (New York: E. P. Dutton & Co., 1953) is among the best known. Chris Bonington, *Everest the Hard Way* (New York: Random House, 1976) and Galen Rowell, *In the Throne Room of the Mountain Gods* (San Francisco: Sierra Club, 1977) are characteristic recent works. A very little article by Lito Tejada-Flores, "Games Climbers Play," *Ascent,* 1964, p. 23, may be the most revealing and remarkable work in the modern literature.

Typical books, illustrative of the participant's view, and striking in both their differences and similarities, are Colin Fletcher, *The Man Who Walked Through Time* (New York: Random House, Vintage Books, 1967), on hiking the Grand Canyon; Joe Henderson, *The Long Run Solution* (Mountain View, Calif.: World Publications, P.O. Box 366, 1976), on running; Sally Wimer, *The Snowmobiler's Companion* (New York: Charles Scribner's Sons, 1973), on snowmobiling; Charles Gaines and George Butler, *Pumping Iron* (New York: Simon and Schuster, 1974), on bodybuilding.

A striking contrast to the Wimer book is Robert Pirsig's *Zen and the Art of Motorcycle Maintenance* (New York: Bantam Books, 1976), a compelling evocation of the possibilities of man and machine in harmony. Of the "real" Zen sports books, the supreme exemplar is Eugen Herrigel, *Zen in the Art of Archery* (New York: Random House, Vintage Books, 1971), closely followed by Michael Murphy, *Golf in the Kingdom* (New York: Dell, Delta Books, 1972). In general, see Alan W. Watts, *The Way of Zen* (New York: Random House, Vintage Books, 1957).

There is a large, serious literature both about sports and "play," a more embracing term whose boundaries are not fully clear. The classic work is Johan Huizinga, *Homo Ludens: A Study of the Play Element in Culture* (London: Routledge & Kegan Paul, 1949). A critique and expansion of Huizinga's thesis is provided by Roger Caillois, *Man, Play and Games* (New York: Free Press, 1961). See also Mihalyi Csikszentmihalyi, *Beyond Boredom and Anxiety* (San Francisco: Jossey-Bass, 1975) and Josef Pieper, *Leisure, the Basis of Culture* (London: Fontana Library, 1965). A glimpse of the extensive psychological and psychoanalytic literature on play may be had in Franz Alexander, "A Contribution to the Theory of Play," *Psychoanalytic Quarterly* 27 (1958):175; and from Rueben Fine, *The Psychology of the Chess Player* (New York: Dover Publications, 1967).

For serious views of organized sports, see Paul Weiss, *Sport: A Philosophic Inquiry* (Carbondale: Southern Illinois University Press, 1969); Allen Guttman, *From Ritual to Record: The Nature of Modern Sports* (New York: Columbia University Press, 1978); and Michael Novak, *The Joy of Sports* (New York: Basic Books, 1976). An interesting historical account is E. Norman Gardiner, *Athletics of the Ancient World* (Oxford: Clarendon Press, 1930).

Mass entertainment and its meaning has been the subject of much writing. Two extensive gatherings of articles are Bernard Rosenberg and David Manning White, eds., *Mass Culture: The Popular Arts in America* (Glencoe, Ill.: Free Press, 1957), and Eric Larrabee and Rolf Meyersohn, eds., *Mass Leisure* (Glencoe, Ill.: Free Press, 1958) (containing a comprehensive bibliography). Two related books worth reading are George A. Pettitt, *Prisoners of Culture* (New York: Charles Scribner's Sons, 1970) and R. M. MacIver, *The Pursuit of Happiness* (New York: Simon & Schuster, 1955). Gilbert Seldes, *The Great Audience* (New York: Viking Press, 1951) offers an unconventional view. Tibor Scitovsky's *The Joyless Economy: An Inquiry into Human Satisfaction and Consumer Dissatisfaction* (New York: Oxford University Press, 1977) is a rare look by an economist into the phenomenon of mass leisure.

The broader psychological implications of the issues raised by the literature on play and on mass culture are explored by Carl Rogers, *On Becoming a Person* (New York: Houghton Mifflin Co., 1961). A more detailed statement of Rogers's views appears in "A Theory of Therapy, Personality, and Interpersonal Relationships, as Developed in the Client-Centered Framework," in *Psychology: A Study of a Science,* ed.

Sigmund Koch, vol. 3 (New York: McGraw Hill, 1959). See also A. H. Maslow, *Motivation and Personality* (New York, Harper & Bros., 1954); idem, *The Farther Reaches of Human Nature* (New York: Viking, 1971); Erich Fromm, *The Sane Society* (New York: Rinehart & Co., 1955); Sigmund Freud, *Group Psychology and the Analysis of the Ego,* trans. and ed. James Strachy (New York: Norton & Co., 1959); Richard Locke, "From TV to Lionel Trilling," *New York Times Book Review,* June 12, 1977, p. 3.

Professional literature on recreation policy is extensive. A highly selected, and representative sample, would include: Marion Clawson and Jack L. Knetch, *Economics of Outdoor Recreation* (Baltimore: Johns Hopkins University Press, 1966); Anthony C. Fisher, *The Economics of Natural Environments* (Baltimore: Johns Hopkins University Press, 1975); Charles J. Cicchetti, Joseph J. Seneca, and Paul Davidson, *The Demand and Supply of Outdoor Recreation* (New Brunswick, N.J.: Rutgers University Press, 1969); B. L. Driver, ed., *Elements of Outdoor Recreation Planning* (Ann Arbor: University of Michigan Press, 1974); John C. Hendee, George H. Stankey, and Robert C. Lucas, *Wilderness Management,* U.S., Department of Agriculture, Forest Service, Miscellaneous Publication no. 1365, October, 1978; *Outdoor Recreation Resources Review Commission, Wilderness and Recreation—A Report on Resources, Values and Problems,* ORRRC Study Report 3 (Washington, D.C.: Government Printing Office, 1962); U.S., Department of Agriculture, Forest Service, *Proceedings: River Recreation Management and Research Symposium,* January 24–27, 1977, Minneapolis, Minn., North Central Forest Experiment Station, General Technical Report NC-28; U.S., Department of the Interior, Bureau of Outdoor Recreation, *Outdoor Recreation: A Legacy for America.* appendix A, "An Economic Analysis" (December, 1973); William E. Shands and Robert G. Healy, *The Lands Nobody Wanted* (Washington, D.C.: Conservation Foundation, 1977); *National Parks for the Future* (Washington, D.C.: Conservation Foundation, 1972); *Toward an Environmental Policy* (Washington, D.C.: National Parks and Conservation Association, 1971); *Preserving Wilderness in Our National Parks* (Washington, D.C.: National Parks and Conservation Association, 1971).

For a view of the preservationist position in the making, the following (selected) periodicals are revealing and informative: *National Parks and Conservation* (National Parks and Conservation Association, 1701 18th St., N.W., Washington, D.C. 20009); *Sierra* (Sierra Club, 530

Bush St., San Francisco, California 94108); *Not Man Apart* (Friends of the Earth, 124 Spear St., San Francisco, California 94105); *The Living Wilderness* and *Wilderness Report* (Wilderness Society, 1901 Pennsylvania Ave., N.W., Washington, D.C. 20006); *High Country News* (331 Main St., Lander, Wyoming 82520).

Index

ROB ANDREW

RUGBY

THE GAME OF MY LIFE

Battling for England in the Professional Era

HODDER

First published in Great Britain in 2017 by Hodder & Stoughton
An Hachette UK company

This paperback edition published in 2018

1

A CIP catalogue record for this title is available from the British Library

B format ISBN 9781473664180
eBook ISBN 9781473664173

Typeset in Scala by Hewer Text UK Ltd, Edinburgh
Printed and bound by CPI Group (UK) Ltd, Croydon, CR0 4YY

Hodder & Stoughton Ltd
Carmelite House
50 Victoria Embankment
London EC4Y 0DZ

www.hodder.co.uk

This book is dedicated to my close school friend Chris McKean, who, despite being paralysed for nearly 40 years, has been a positive and supportive presence in my life since the time we spent together on the rugby and cricket fields of Barnard Castle. A very special person.

I would also like to add my thanks and gratitude to Chris Hewett and David Norrie for all their support, help and humour in putting this book together and to my publisher Roddy Bloomfield, Fiona Rose and all their colleagues at Hodder for making this possible. I hope you enjoy the story of my journey.

CONTENTS

TIMELINE

1995

England win the Five Nations Grand Slam.

Will Carling is sacked as captain for describing Rugby Football Union members as 'old farts', then reinstated after an uproar.

England overcome the reigning champions Australia in a dramatic World Cup quarter-final, thanks to Rob Andrew's late drop goal, but are then heavily beaten by a Jonah Lomu-inspired New Zealand. The tournament is awash with rumours about a professional 'breakaway' competition involving the sport's leading players.

The International Board declare the sport 'open' following Rupert Murdoch's buy-up of the game in the southern hemisphere. Twickenham declares a one-year moratorium, so the game in England remains amateur, if only in theory. Rob Andrew leaves Wasps for a full-time role at second division Newcastle and calls time on his international career.

1996

England retain the Five Nations title, despite losing to France in Paris.

Rob Andrew accelerates his team-building process at Newcastle in a bid to guarantee promotion to the top flight.

1997

England slip to second in the Five Nations following a home defeat by the Grand Slam-winning French. Rob Andrew makes a surprise farewell international appearance off the bench in the final game against Wales. Jack Rowell is succeeded as head coach by Clive Woodward.

The British and Irish Lions win a compelling Test series in South Africa.

Newcastle win promotion to the first division, finishing second to Richmond in the Division Two title race.

1998

England again finish second to France in the Five Nations and are then humiliated in Australia and New Zealand on the summer 'tour of hell'.

Francis Baron becomes the RFU's first chief executive.

Newcastle win the inaugural Premiership title. Rob Andrew plays every game. The English clubs announce a boycott of European rugby in the 1998/99 season following a row over the running of the Heineken Cup.

1999

England finish second to Scotland in the Five Nations and go on to lose to South Africa at the quarter-final stage of the World Cup.

Newcastle finish eighth in the Premiership. Rob Andrew retires from playing following an early-season injury in training. Begins work on the 'Andrew Plan' in an attempt to stabilise a politics-riven domestic game.

2000

The Five Nations becomes Six with the inclusion of Italy, and England win the title, despite losing their final game in Scotland.

Newcastle finish ninth in the Premiership. The 'Andrew Report' fails to win the approval of the RFU council and is not implemented.

2001

England retain their Six Nations title but fail to win the Grand Slam, losing to Ireland in a finale delayed by an outbreak of foot and mouth disease.

The British and Irish Lions lose a close Test series in Australia.

Newcastle finish sixth in the Premiership and win the English knock-out cup, beating Harlequins 30-27 in the final at Twickenham.

2002

England finish second in the Six Nations, behind the Grand Slam-winning French. They beat all three southern hemisphere superpowers in the autumn series at Twickenham – the All Blacks by three points, the Wallabies by one and the Springboks by 50.

Newcastle hold their sixth-place position in the Premiership.

2003

England win the Six Nations Grand Slam, a first clean sweep since 1995. They beat New Zealand and Australia in Wellington and Melbourne respectively and go on to triumph at the World Cup, recording a famous extra-time victory over the Wallabies in the final.

Newcastle finish tenth in the Premiership.

2004

England drop to third in the Six Nations, losing to both Ireland and France. Clive Woodward resigns as head coach following a disappointing summer trip to the southern hemisphere and is replaced by his assistant, Andy Robinson.

Newcastle finish ninth in the Premiership but reclaim the knock-out cup, beating Sale in the final at Twickenham.

2005

England finish fourth in the Six Nations, winning only two games.

The British and Irish Lions lose all three Tests in New Zealand – their worst performance since 1983.

Newcastle finish seventh in the Premiership and make it through to the knock-out stage of the Heineken Cup, losing in the quarter-finals to Stade Francais.

Martyn Thomas is elected RFU chairman.

2006

England again finish fourth in the Six Nations. Robinson remains in charge, but the RFU dispenses with his coaching team.

Newcastle finish seventh in the Premiership. Rob Andrew leaves the club after almost 11 years to take up a new elite rugby director role at Twickenham, beating Clive Woodward to the job.

England lose three of their four autumn internationals. Robinson is sacked, with Brian Ashton taking over the head coach position.

2007

England rise to third in the Six Nations, but then send a weakened side to South Africa and are badly beaten in both Tests. Against expectations, they reach the final of the World Cup in France, eventually losing to the Springboks.

Rob Andrew leads the RFU team in securing a far-reaching agreement with the top-flight clubs over player release and finance.

2008

England finish second to Grand Slam-winning Wales in the Six Nations, but Ashton loses his job despite the improved performance. Martin Johnson, the former England and Lions captain, is named as his replacement.

In Johnson's temporary absence, Rob Andrew leads England on an unsuccessful two-Test summer tour of New Zealand, during which a number of players are accused of sexual misconduct. No charges are brought.

2009

England are again runners-up in the Six Nations, beaten by the Grand Slam-winning Irish.

The British and Irish Lions lose a tight series against the Springboks in South Africa.

2010

England slip to third in the Six Nations, winning only their first two games, but perform well to square a summer series in Australia. However, they lose to both New Zealand and South Africa at Twickenham in the autumn.

Francis Baron steps down as RFU chief executive. John Steele, a former UK Sport CEO, replaces him.

2011

England win the Six Nations, despite losing their final game in Dublin. During a World Cup campaign undermined by disciplinary problems, they struggle to qualify for the knock-out stage and are duly beaten by France in the last eight. Details of a confidential RFU report into events at the tournament are leaked to the press. Martin Johnson resigns. Stuart Lancaster is appointed in his stead.

John Steele's brief tenure as CEO comes to an abrupt end, sparking political turmoil at Twickenham. Martyn Thomas takes over in a caretaker capacity, but is gone by the end of the year. Ian Ritchie takes over the top job.

2012

England surprise most observers by finishing second to Wales in the Six Nations. They also perform well on the summer tour of South Africa, drawing the last Test in Port Elizabeth. In the autumn, they record a remarkable victory over the world champion All Blacks at Twickenham.

2013

England challenge hard for the Six Nations title but finish second to Wales after an emphatic defeat in Cardiff.

The British and Irish Lions win their Test series against Australia.

2014

England again miss out on the Six Nations title, beaten to the prize by Ireland. They go close to beating the All Blacks in Auckland in the opening Test of their summer tour but end up losing the series 3-0. They are also defeated twice at Twickenham during the autumn.

2015

A third successive second place for England in the Six Nations: their high-scoring victory over France in the final match is not enough to overhaul Ireland. Hopes are high for the World Cup campaign on home soil, but England become the first host nation to exit the tournament at the group stage. Lancaster and his coaching staff leave their posts. Eddie Jones, an Australian, becomes the first overseas head coach of the national team.

2016

England win the Six Nations Grand Slam and also break new ground by winning a Test series in Australia.

Rob Andrew secures a second long-term agreement with the Premiership clubs. He announces his departure from the RFU after a ten-year stint at Twickenham and subsequently joins Sussex County Cricket Club as chief executive.

1
DROP GOAL DÉJÀ VU

IT WILL ALWAYS be with us, I suppose. Thanks to the wonders of the Internet, you are never more than a mouse-click away from hearing BBC Radio's live account of the build-up to, and execution of, Jonny Wilkinson's drop goal against Australia in 2003 – the kick that made England champions of the world for the first time and propelled Jonny into the distant galaxy of rugby celebrity that had previously been the exclusive preserve of a certain outsized All Black wing by the name of Jonah Lomu. The question is this: what was 'it', exactly?

There is no mystery surrounding Ian Robertson's role in the broadcast. With less than a minute of extra time left on the clock in Sydney, the longest-serving rugby commentator of them all knew precisely what England had in mind as they attacked off that final line-out, and he successfully captured the wrenching agony and eventual ecstasy of the denouement, even if Matt Dawson and Martin Johnson conspired to wrong-foot him a little by going off-script in an attempt to give Jonny a better sight of the sticks than he might otherwise have had. Me? I said nothing at all. Not, at least, in words. My contribution was outside the confines of formal language: it was simply a sound – two, actually, although most people remember only the first – that bordered on the primal and would surely have been recognised by the cavemen of old, including the Palaeolithics who played in the front row during the decade or so I spent in the England shirt Jonny had inherited. And while I've often wondered whether I ruined Ian's verbal work of art with my involuntary and unprofessional partisan noises – the equivalent of a smudged thumbprint on an otherwise perfect oil painting, or the dropping of a cymbal during the slow movement of a symphony – I can't remember ever speaking to anyone who shared that view. All things considered, it seemed to go down rather well.

Some people thought they heard a long 'yeeeessss' as the ball bisected the posts, leaving the Wallabies nowhere to go but down on their knees in despair. Others have described it as a 'yell', a 'yelp', a 'shriek', a 'scream', a 'squeal', a 'howl'. On reflection, I don't think any of those words quite fit the bill. To me, it was . . . an outpouring. An outpouring that rose from somewhere deep down in my sporting soul and gave voice to a rugby lifetime's worth of experiences, good and bad.

The most obvious reference point was the quarter-final of the World Cup in South Africa in 1995: another do-or-die meeting between England and Australia; another hellishly tense contest decided in the last seconds by a drop kick from a No. 10 clad in white. On that occasion, I had been the one burdened with the responsibility of performing the task rather than passing judgement on it, and to this day, I'm spooked by the similarities. Same opposition, same pressure of time, the same line-out launch pad and the same inescapable fact that, like Jonny, I would have only a single shot at it. One chance to get it right.

It had been a heck of a scrap, down there in Cape Town. The Wallabies were the reigning champions and were probably the favourites for the title, even though they'd lost to South Africa in the opening match of the tournament. Some of the players in their starting line-up that day – David Campese, Jason Little, Tim Horan, Michael Lynagh, Phil Kearns, John Eales, Willie Ofahengaue – had already established themselves as all-time Wallaby greats, while others – Matt Burke, George Gregan – would do so in the years that followed. We, meanwhile, rather fancied our chances of going the full distance to the final if we could somehow find a way to win this one. Lynagh had been in the zone with his kicking all afternoon, as had I, when, well into injury time, we were awarded a penalty deep in our own half. (As I remember it, Campese had taken it upon himself to attempt a drop goal of his own – he probably shoved the rather more proficient Lynagh out of the way to do it, knowing him – and when it went horribly wrong, we were finally able to lift the Australian siege and work ourselves into a safer part of the field.) On the way to the line-out, I said to our hooker, Brian Moore, and our skipper, Will Carling: 'Look, I'll have a crack at this if we can just get the ball clean and drive it upfield a little bit.' We didn't have a heavily rehearsed, intricately choreographed move of the kind we would see eight years later in Sydney; just a basic idea of what might be worth trying. Sure enough, Martin Bayfield delivered quality possession from Brian's throw and the pack shunted the Wallabies back a good few metres, tying the Australian flanker David Wilson into the maul in the process. I was still a long way out – the best part of 50 metres, including the angle – but without the incarcerated Wilson hunting me down, I at least had some room. And the moment I struck it, I knew. I'd nailed a fair few drop goals for England over the course of a decade's international rugby, but this was the one and only time in my life when I thought: '*Jesus!* That's gone miles. That's in outer space.' I couldn't have hit it any more sweetly if I'd stayed on the Newlands pitch for another month. It was still going up as it went over.

There was a feeling of elation, of course, but when these things happen, the emotional reaction after the initial buzz is far more difficult to define. I'm sure Jonny feels the same way – and heaven knows, he's spent more time than most trying to rationalise his responses to things that happen in the sporting arena. It's a mix of many ingredients. Pure joy plays no bigger part than sheer relief – the profound sense of satisfaction and, I admit it, self-justification, is often invaded by memories of your many failures in the past and the nagging fear that you might mess up in the future. You spend your life practising this stuff, hour after hour after hour, because you know the last-gasp drop goal is such an essential part of an outside-half's armoury. In a sense, it is rugby's version of the nuclear option, the ultimate 'cometh the moment' weapon: you get to press the button once and once only, with everything hanging on the outcome. It is not a matter of hope or faith, but one of belief. If you don't believe, right down there at the core of your spirit, that you've mastered the relevant technical demands and hard-wired yourself into a state where you can shift to autopilot at the crucial moment and deliver something in front of a mass audience that you routinely deliver in training when the stadium is empty, you'll miss. You find out an awful lot about yourself, dropping goals at the death.

Yet there was a great deal more behind my reaction in Sydney than a simple kicking comparison. For one thing, I didn't fully understand until then how deep the England–Australia sporting rivalry runs. Years of reflection have left me convinced that in rugby union – and, perhaps even more so, in the great game of cricket – there is something unique about this particular struggle for supremacy. I do not think I'm being unreasonable, still less smug or overbearingly English, in suggesting that the competitive juices on the far side of the world are not flowing as strongly as they once did – that the Aussie brashness of old has become a little blunted at international level. It seems to me that the startling growth of Australian Rules Football and the country's continuing dominance in rugby league has led them at least some of the way down the American road of prioritising domestic sport at the expense of efforts in the international arena. But the old anger-fuelled determination will return soon enough, particularly when it comes to beating the Old Country – and, when it does, watch out. I don't think I've ever been as shaken as a spectator as when, soon after flying to Australia in 1986 for a spell of rugby in Sydney with the Gordon club, I was taken to a State of Origin league game. I was transfixed by what I saw. More than that, I was genuinely shocked by the fury, the intensity, the absolute brutality of it. That memory returned as Jonny landed his kick more than a decade and a half later and played its part in my outpouring.

There was also something of the personal relationship I'd built up with Jonny during our years together at Newcastle Falcons: player to player, teacher to student, boss to employee, old head to young spirit. I knew what he'd been through to earn his place in that England team and I understood exactly what battles he had been fighting, psychologically and emotionally, during the tournament. To say I was proud of him would be the understatement to end them all, just as I was proud of the team as a whole, for I'd played alongside a good number of Jonny's colleagues in the squad and against every one of them. To take just one example, I could remember Lawrence Dallaglio walking into Wasps as a 17-year-old and telling the senior players of the time – myself, Steve Bates and the fearsome Dean Ryan, among others – that he should be in the starting team instead of . . . Dean Ryan. And in the team as captain, too. A confident sort, our Lawrence.

Lastly, there was the wonderful realisation that England had finally won the global title and that, while my own attempts had ended in failure, this was the next best thing. Those failures had been painful. Winning a World Cup is the ultimate. It's phenomenally difficult just to get close to it, and when you lose a final, as we did against Australia on home territory at Twickenham in 1991, or find yourselves being blown out of a semi-final in an unexpected and unusually soul-sapping fashion, as we did in South Africa four years later, that's something you never get back, a hole you never fill. I'm not sure I've ever put that 1991 defeat behind me. Not fully. It was a 'could have, should have, didn't' moment: the opportunity of a lifetime, spurned for a lifetime. I would not go as far as to say that we went into that competition expecting to win the title, but after beating the French in Paris and the Scots in Edinburgh to make it through to the decider, we certainly believed we were equipped to see off the Wallabies, irrespective of the hard-won experience of Lynagh and Nick Farr-Jones at half-back and the rich potential of Little, Horan, Kearns and Eales. And we probably would have won, had Campese not denied Rory Underwood a run-in try with a knock-on so spectacularly cynical that he might spend an entire season in the sin bin if he pulled the same stunt today, and had Eales not pulled off a miracle tackle on yours truly as I headed for the line late on. (Well, I never claimed to be the quickest.)

On a different day the trophy would have been ours and,

had it worked out for us, I think that England side would now be seen in the front rank of World Cup-winning teams, if not quite as good as the 1999 Wallabies or the All Blacks who won in 1987 and 2015. But it wasn't a different day. It was *that* day. And that day hurts me still. Like a batsman who is closing in on a century after hours of painstaking concentration and attention to detail and then nicks one behind – and knows he's nicked it, yet still stands there in the blind hope that the keeper has dropped it or the umpire hasn't heard it, even though his heart is telling him neither thing is remotely possible – the misery of our missed opportunity at Twickenham laid me low.

As a player, you do not always register the full enormity of an occasion: for instance, I spent a good deal of time reflecting on the tumultuous events of the 1995 World Cup before I even began to process what had happened, not just in terms of the rugby on the field and the ground-shifting moves towards professionalising the sport that were taking place off it, but also with regards to the political upheaval of a nation in ferment, with Nelson Mandela so recently installed as the country's first non-white president. On the flipside, there are moments in time that freeze in your mind: moments when you know a chance has gone and won't come again. Some things in life are irretrievable. I can think of only one player – my old colleague Jason Leonard, the grand old England front-rower – who has won a World Cup final after losing one. For the rest of us, it is a question of living with the consequences of coming up short and finding salvation in the glory of your successors.

I don't sit around beating myself up about the 1991 final: so much has happened to rugby, and to my life in rugby, in the intervening years, I sometimes think that game was played while dinosaurs still roamed the earth. In a manner of speaking, that was indeed the case: there were more than a few brontosauruses lurking in the Rugby Football Union council room, as we were all about to discover. But you play sport because you want to achieve something when your time arrives, and our time came and went without us managing to close the deal. As a result, my most joyful moment in a World Cup final was not my moment at all. It was Jonny Wilkinson's. Hence the outpouring.

Of course, I've since pointed out to Jonny that his drop goal wasn't much to write home about: that it was nothing more than a schoolkid's effort – short range, in front of the sticks – rather than a proper grown-up job from the best part of 50 metres, kicked on the diagonal. But for the fact that Jonny nailed his shot off the wrong foot, in the last 20 seconds of extra time in a World Cup final against Australia, in Australia, with the Webb Ellis Trophy sitting there on the touchline waiting to be held aloft, it would have had nothing to recommend it at all! Okay, so I'm clutching at straws.

I am struck by one other parallel between my Cape Town success and Jonny's high point in Sydney. Back in 1995, our elation lasted precisely a week: the time it took us to spend a couple of days on the celebratory sauce, bend our minds towards the semi-final against New Zealand and then run smack-bang into Jonah and his friends. If the 2003-ers were granted a little longer to enjoy the spoils of victory, the feel-good factor did not last beyond the end of the European season. By the time the 2004–05 campaign began, England had yielded their Six Nations title to France, been spanked by both New Zealand and Australia on the summer tour, lost Wilkinson to long-term injury, suffered the shock of Clive Woodward's resignation as head coach after a seven-year stretch, and heard the influential Dallaglio announce his retirement from international rugby, albeit a retirement of the temporary variety. It was almost as if the biggest union-playing nation on the planet was too fractured, too wrapped up in its own internal squabbles and contradictions – and, crucially, too leaderless – to build on the achievements of its shop-window team. Which it was. Did this come as a surprise to me, as I sat in my office on Tyneside preparing Newcastle Falcons for another long slog through the Premiership? Not in the slightest. I had seen it all before, more than once.

2

THE WORLD TURNED UPSIDE DOWN

LIKE MOST FRONT-LINE sportsmen, international rugby players spend a lot of their time – maybe too much of their time – thinking in micro rather than macro. What happens tomorrow, or in next week's match, is the only thing that matters. Next year, the next five years, the long-term prospects for the sport – all that can take care of itself. Those of us involved in the 1995 World Cup did not, however, require a professorship in hindsight to recognise that certain things were happening that would change the game in the most fundamental of ways. For a start, there was Jonah Lomu, an inexperienced All Black of Tongan descent who, in the space of a single 80-minute performance, brought the union code to a whole new audience and alerted businessmen and television executives from all four corners of the earth to an untapped income stream of considerable potential. In fact, it took him rather less than 80 minutes, for the contest he chose as his showcase was pretty much over in a quarter of that time. Unfortunately for me and the rest of an England team just beginning to think in terms of a second successive global final, we were the fall guys. I'd never seen any player make such an impact before our meeting with New Zealand, and I don't think for a moment that I'll see anything like it again. In that sense, if in no other, it was a privilege to be on the same pitch.

Above and beyond that most celebrated of individual displays – and yes, there was something bigger than Jonah going on in South Africa just then, hard though it is to believe – was the cloak-and-dagger stuff taking place in hotels and restaurants in every major city in the country, from Durban and Cape Town on the shores of different seas to Pretoria and Johannesburg up there on the Highveld. The matter under discussion was a breakaway rugby competition that would not merely feature the world's best players – not just some of them, but virtually all of them – but also propel the sport headlong into the professional age.

The man behind the project was Kerry Packer, the Australian media tycoon who had trodden a similar path in cricket almost two decades previously. Had rugby gone the same way, and believe me, there were plenty of us willing to throw our weight behind it, there would have been an even greater upheaval. When Packer launched his World Series Cricket venture in 1977, his chosen sport was already run on a pay-for-play basis, even if the people providing the entertainment were being paid precious little by the governing bodies. In 1995, the England rugby players were still fitting their training in and around their working lives while fighting tooth and nail with the Twickenham authorities over matters of interplanetary significance, such as travel expenses. At the time, we were receiving the princely sum of 15 pence a mile, with the threat of being docked £1.50 if someone suspected us of committing the heinous crime of over-calculating the travelling distance for personal gain.

Those of us in what would these days be called the 'senior player group' – Will Carling as captain, Brian Moore as provocateur-in-chief and myself, along with the ultra-phlegmatic Dean Richards, who generally went with the flow but sometimes felt sufficiently strongly about an issue to make his feelings known – had fought the odd full-scale battle and a series of small skirmishes with the union over commercial issues during the early 1990s, and those of us prepared to give proper thought to the future of the game knew perfectly well that professionalism in some form or other was inevitable. But by the time the 1995 World Cup came around, our relations with Twickenham were in a state of permafrost: the atmosphere was ice-cold whenever the subject of money cropped up. It would take a truly significant development to heat things up and bring them to boiling point. Something like the offer of £100,000 a year to play in the rugby union version of World Series Cricket.

When we travelled to South Africa, we had already heard whispers of a big-money breakaway project, and as we were already in unusually militant mood, we were more than willing to listen to whatever proposal might be on the table. In pure rugby terms, we were in a good place as a squad: the transition from 1991 had been relatively trouble-free by England standards – there was none of the roll-the-dice selection nonsense I'd experienced in my early years as an international player back in the mid-1980s. If, on reflection, I think the 1991 side was a little more accomplished than the 1995 version, some of the newcomers were of a very high quality indeed: Martin Johnson, Tim Rodber, Ben Clarke; these were players who could handle themselves under Test-match pressure at its most extreme. Geoff Cooke, the head coach who had spent seven years making us competitive, had walked away in 1994, but with Jack Rowell moving from the wildly successful Bath club to replace him, we immediately squared a two-Test series with the Springboks in South Africa before winning the Five Nations Grand Slam for the third time in five years. As that was our last act before the World Cup, why wouldn't we have felt confident?

Things were not quite as rosy off the field, however. During that Springbok summer of 1994, the South African Rugby Football Union president Louis Luyt – a man who preferred to use the blunt instrument rather than the sharp-edged sabre when it came to public speaking – made it abundantly clear during one of his post-match addresses that if the game had not turned professional before his country hosted the World Cup in a little under a year's time, it would do so pretty damned quickly afterwards. Among those listening were the likes of the RFU president Ian Beer, a die-hard rugby traditionalist who had been headmaster at Harrow School, and Dudley Wood, the highly intelligent and articulate secretary of the union who, for all his abilities, could not reasonably be described as one of life's natural radicals. However we, the mere underlings, were unsurprised by Luyt's comments, because we knew full well that the leading South African players were already tapping into the riches that lay beneath the surface of the sport. The New Zealanders were doing the same, as had become obvious to anyone with a functioning pair of eyes during the Lions tour of that country in 1993, but our lords and masters did not see it the same way, and therefore refused to take even the slightest notice. To be fair to Don Rutherford, the RFU's long-serving technical director, the moment was not lost on him. Don was more of a moderniser than most people understood, and I felt he always gave us a fair hearing when we tried to push the envelope on the commercial front. But in the end, he was a union employee – a high-level employee, but an employee all the same. He could not sway the argument; he didn't have the muscle.

So there we were, not so much off the pace set by our southern-hemisphere rivals as stuck in the starting blocks. We had achieved hardly anything over the course of our discussions with the RFU in recent seasons. Four years previously, I'd been dragged in front of the governing body's council along with Will and Brian because we were deemed to have been the ringleaders of the players' commercial campaign leading into the 1991 World Cup. Which we were. Will might not have been everyone's cup of tea, but as captain he was completely committed to doing his best for the people playing under him, and saw it as his job to drive things every bit as hard outside the 80 minutes on a weekend as he did on match day. Brian was just Brian: he loved a scrap and relished nothing more than 12 rounds of niggle with the high-and-mighty types at the union. He had very strong ideas on a range of subjects; if he could poke a bear or, rather, bash a bear over the head, he couldn't help himself. Me? I just enjoyed sticking my oar in. I'd been around the England scene for a good while, I'd thought quite carefully about the way things were unfolding across the sport worldwide, and I was always keen to push the boundaries. It's a part of my make-up: if something intrigues me enough, I go with it completely because I want to know where it will take me. So it was that, in 1991, I found myself in the less-than-glamorous position of honorary treasurer for the England players' 'Run With The Ball' campaign, which we set up in an effort to generate some money from off-field sponsorship activity. When we appeared before the council to explain ourselves, we'd raised the mighty sum of £26,000 before tax, which would be divided between 26 players. We'd done our best on the financial front, but by comparison with what was going on south of the Equator, it was a bad joke, as was the fact that I had been the one pictured 'running with the ball' to promote the venture. As more than one of my colleagues pointed out, it didn't quite fit my image.

With no sign of a thaw between players and union as we approached the big tournament in South Africa, frustration was beginning to build. There was no open animosity, more a game of cat and mouse, but some of us were getting seriously frustrated. We knew the status quo was unsustainable but, try as we might, we couldn't persuade the really important people at Twickenham of that. And the older members of the squad, myself included, were uncomfortably aware that having worked so hard to find a way out of the doldrums of the mid-1980s and close in on the All Blacks and the Wallabies, we were about to fall behind them once again. It was against that background that Will made his '57 old farts' comment in respect of the Twickenham top brass.

You could call it a comedy of errors – not that Will would spend much of his time laughing as the ensuing farce unfolded. He had given a television interview to Channel 4 – Greg Dyke, who would go on to make a serious name for himself in the football world as well as the broadcasting one, was the man with the microphone – and he had made his remark as he left the room, unaware that he was still being recorded. He could have used any number of descriptions: 'dusty relics', maybe, or 'crumbling waxworks'. But he went for the elderly–noxious combination, and all hell broke loose the moment the tape was leaked. A matter of days before the World Cup, the RFU stripped him of the captaincy. Brilliant.

I don't think any of us thought for a second that he would be sacked. I've never made a detailed study of it, but I suspect that the reaction of most of the England players on hearing the 'old farts' comment was to think: 'That's a fair reflection.' It certainly resonated with me. When news of the captaincy decision broke, on the morning of the Pilkington Cup final in which I was playing for Wasps against Bath, there was a flurry of phone calls between various members of the squad. We were rock-solid in agreeing that anyone approached to take on the role of skipper would instantly knock it back. We made sure that message reached both Jack Rowell, who was keeping his head well down, and Don Rutherford, who had full access to the high command at Twickenham. What was more, we were prepared to go a step further. Our position was as hard line as it was possible to imagine: 'Either you reinstate Will,' we said, using back channels we knew led straight to the council room, 'or you can find yourselves a whole new squad for the World Cup. As things stand, we won't be going to South Africa.'

Given the scale of Will's profile as a successful England captain (he was big news in those days), there was precious little chance of Dennis Easby, the RFU president, winning the hearts and minds of the sporting public. Easby must have realised that much from the wall-to-wall press coverage, which was a long way short of flattering. I felt sorry for him, if I'm honest. Here was a man in his 70s, a retired solicitor from Reading, copping it full-bore from every direction over a decision that had almost certainly been taken by other people – by a cabal of far more influential grandees, many of whom considered the players to be too uppity by half. It was a monumental cock-up by the union and I imagine Easby knew it. He must have been seriously relieved when, in next to no time, there was a stage-managed rapprochement with Will during a squad session in Marlow: smiles, handshakes, photographers by the dozen. 'Right,' we thought, 'we've won that one. Off to South Africa we go.' It had been a bizarre episode, to put it mildly.

If none of us had foreseen the mini-crisis over the captaincy, we were even less prepared for the extraordinary events that unfolded in Springbok country over a few short weeks in May and June. The rugby itself was something else: even though we were pretty sluggish through the pool stage, we ended up featuring in two of the most dramatic knock-out matches in the game's modern history. By losing to South Africa in Cape Town in the opening match of the tournament, the Australians changed the entire shape of the draw, not least for us. Confident that we would top our own group, which was probably the least hazardous of the four, we rather assumed that the Wallabies would be too good for the Boks and that we'd play the hosts in the last eight – no easy matter, by a long chalk, but doable. We had, after all, trounced them in Pretoria just a year previously. As it turned out, we found ourselves facing the Wallabies in the quarters . . . and then the All Blacks.

Strange as it may seem in the light of the New Zealanders' phenomenal performance in recent years (only a fool would attempt to deny that they have set the benchmark in terms of the skill levels of their players, the quality of their attacking rugby, their matchless continuity in selection and their mastery of the big-match mentality), that 1995 team was something of an unknown quantity. Yes, there were some familiar faces in the side: anyone who knew anything about rugby recognised the threat posed by Frank Bunce in the centre, Sean Fitzpatrick at hooker and the multifaceted Zinzan Brooke at No. 8. But what did we know of Glen Osborne or Andrew Mehrtens or Josh Kronfeld? As for this Lomu bloke on the wing – he was big, apparently, but was he any good? The gaps in our knowledge would soon be filled, sadly.

Looking back, I believe our preparation for the semi-final was far more wrong than right. After beating the Wallabies in Cape Town and celebrating in enthusiastic fashion, we continued along the path of liquid relaxation by spending a couple of days in the resort of Sun City, north-west of Johannesburg. Quite why we went, I'm not sure: all I know now is that the decision to fly from sea level to altitude for a spell of brash, in-your-face downtime – if the trip was about rest and recuperation, there was very little of either – was some way short of inspired. As a consequence, we were simply not set up psychologically to maximise our chances over the remainder of the tournament. Even in 1995, when the great advances in sports science with which we have become so familiar had yet to be made, success in international sport was a matter of small margins. When we took to the field against the All Blacks, those margins were dominated by one very big bloke.

Rugby union has had its share of incredible individual performances down the years, but Lomu's display against us was beyond ridiculous. We contributed to our own downfall by being naïve to the point of absurdity. We were still in the days of chatting about defence rather than practising it – of deciding whether to drift off scrums and line-outs or go for the man-for-man option, hitting a few tackle pads at the end of training, holding the same pads for someone else and then heading for the shower feeling properly organised – and it was on this meagre basis that we went into the game believing we had Jonah 'worked out'. If he'd looked impressive enough on the television against the three Celtic nations (in fact, he'd ripped them up), we were far from convinced that any of those teams could tackle the way we tackled. Then, within a few seconds of kick-off on the big day, I found myself thinking: 'Oh . . . My . . . God. He's messed us up straight from the kick-off, he's whacked Will out of the way, Tony Underwood is in Row Z, Mike Catt has been buried in a shallow grave and . . . yep . . . he's scored. Mmmm. We may need a rethink here.'

Minutes previously, I'd been balancing my suspicion that we weren't as well prepared as we should have been against the knowledge that in 1993 I'd been on the winning side against this lot twice in a few months. Now, I had a sick feeling in the pit of my stomach: the feeling you get when you know that, however this game pans out, it's unlikely to end well. Sure enough, it was over as a contest by the end of the first quarter. Generally speaking, top players react positively to adversity: the competitive edge kicks in hard and forces you to believe there's still something to be taken from a contest, even when you're being splattered all over the field. But there are times when you know, in your heart of hearts, that this belief is shallow at best and delusional at worst. Deep down, there's a feeling of helplessness, of something close to futility. I would never have admitted to it to anyone during or immediately after the semi-final, but the feeling was there that day. The fact that we won the second half remains one of the most pointless facts in the whole of sport.

Our reward for not handling Jonah in quite the way we'd imagined was a third-place play-off against France in Pretoria – a fixture neither team wanted to play, and one that was not so much overshadowed as obliterated by the increasingly frenzied chatter about Kerry Packer's money and the proposed breakaway. As England players we had been heavily engaged with the project almost from the beginning, with contacts made through Packer's man on the ground, the one-time Wallaby prop Ross Turnbull. I'd met him more than a decade before the tournament: he was a friend of Alan Jones, who had coached a genuinely great Wallaby side on their Grand Slam-winning tour of the British Isles in 1984, and with whom I subsequently spent time during my brief spell of club rugby in Sydney. I wasn't quite sure at that point what role Ross played in rugby, but he was clearly tuned into the business side of the sport, and I wasn't wholly surprised to find him at the heart of things in South Africa. He was one of those go-get-it entrepreneurial types who would shake the English club game to its roots in the newly professionalised era, and he certainly talked a good game as he spent the World Cup visiting the hotels of the major contenders – England and France, the three southern-hemisphere superpowers – and set about dangling the promise of untold riches in front of the senior players, who promptly fed the relevant information back to their fellow countrymen.

The plan was to sign up the vast majority of the five leading squads, if not snap up those squads lock, stock and barrel, for a series of televised fixtures to be hosted in cities across the world, including top venues in the United States as well as in the established union heartlands. If he could collect enough signatures, so went the theory, a domino effect would be created: those who had signed would return to their own countries and persuade others to take part. It was a highly ambitious plan, apparently with finance to match. Six-figure sums were routinely mentioned, and the talk dominated conversation among the players, the overwhelming majority of whom were more than a little interested in securing a piece of the action.

Naturally, a degree of healthy scepticism could be detected among the players: there was never a feeling of complete certainty that the money truly existed, despite Packer's high-profile history in the field of sporting upheaval. But there was also a widespread willingness to engage with Turnbull, fuelled by the twin elements of curiosity and exasperation. The curiosity was not simply a money-driven thing: the World Rugby Corporation, to use its working title, was a radical idea from top to bottom, and for those of us who had been pondering the future direction of the game we loved, the prospect of a completely new departure was inherently exciting. The exasperation, at least from the point of view of the England contingent, was entirely driven by our failure to drag our so-called lords and masters at Twickenham out of the feudal age. Whatever progress we had made – and there wasn't much of it – had been so incremental as to border on the meaningless.

Four years previously, I had spent some time playing club rugby in France with Toulouse, and while I still had a day job in surveying with the sister company of my employers in London, I trained four times a week rather than two and received a monthly expenses provision, extremely handsome by the standards of European rugby union at the time, for my trouble. To make matters more interesting still, I was paid through the Toulouse payroll. Forget the old brown envelopes stuffed with cash. This was a brown envelope with a payslip, and it accounted for a decent chunk of my overall income. When I returned to England to pick up where I'd left off with Wasps, it was a case of status quo ante. Nothing had changed since I'd been away and there was no indication from Twickenham of any appetite for change in the future.

To have gone from there to a £100,000 contract would have been game-changing in all senses of the phrase, and pretty much every member of the England squad in South Africa recognised that fact. One or two of our party with major ambitions away from rugby might have hesitated on the grounds that they had lucrative careers to consider, but had Turnbull slapped a contract on the table and said 'sign here', I remain convinced to this day that the overwhelming majority would have replied: 'Where's the pen?' The same feeling was running through all the top-ranking teams in the tournament. That much became abundantly clear when we arrived in Pretoria for the play-off game we really didn't want to play.

There were only four sides left by that time: ourselves and the French, faced with the less-than-tantalising prospect of a meaningless fixture before a semi-interested audience at the great Loftus Versfeld Stadium, which was an immeasurably better venue than the match it was staging; and the two finalists, South Africa and New Zealand, who would be taking centre stage at Ellis Park in downtown Johannesburg a couple of days later. It would be a significant overstatement to suggest that we were relishing our meeting with Les Bleus: apart from anything else, there was no common ground between us. Not to put too fine a point on it, we'd hated them for years. I felt I had more understanding of what they amounted to than some of my colleagues, having spent a chunk of time on their side of the water. (Not that my attempts to improve cross-Channel relationships saved me from receiving a boot in the face during the infamous Five Nations game in which two of the French front-row forwards, Grégoire Lascubé and Vincent Moscato, were sent off. While I like to think it was accidental – after all, I was playing for Toulouse at the time – it still hurt.) But, generally speaking, there was no relationship between the two sets of players. Brian Moore had dedicated a good part of his career to stoking the fires of Anglo–French enmity, and his was the majority view in the England dressing room.

Imagine how we felt when, having checked into our hotel in Pretoria, we decided to ease our misery by heading for the bar around the corner, only to find that the French boys had beaten us to it. We were tired, having been run off our feet by Jonah and his mates a few hours earlier, and we were thoroughly hacked off with life. And now, by way of adding insult to injury, we couldn't even have a drink. As we stood there, looking at our next opponents as they were looking at us, in a kind of Mexican stand-off, somebody said: 'Bugger it, we're staying.' A couple of drinks later, we were standing at the bar together with our rivals singing 'La Marseillaise' at the tops of our voices. That was the Sunday night. We did it again on the Monday night, and again on the Tuesday night. By the time the tournament ended, we were the best of buddies. Relations have been comparatively stable ever since.

Needless to say, much of the banter between us concerned the breakaway proposals. Philippe Saint-André, the French captain, had been high on Turnbull's target list for preliminary discussions and it quickly became clear to me that lots of his teammates were every bit as enthusiastic as us, if not more so. Their problems with the French Federation were not dissimilar to ours with the Rugby Football Union. In fact, they may have been worse. When we weren't singing at the bar, we were talking quietly among ourselves in an attempt to make sense of what was happening. 'Are you blokes going with Turnbull?' someone would ask. 'Absolutely. Are you?' came the reply.

Having established that the Wallabies were also on board – we knew that much before we put them on the plane home to Sydney – and having satisfied ourselves that Francois Pienaar and Sean Fitzpatrick, the Springbok and All Black captains respectively, were among those driving the project forward, we were eagerly awaiting the appearance of a piece of paper that would set events in motion. We were not alone. As losing semi-finalists, both the French and ourselves attended the Ellis Park decider and then headed to the outskirts of the city for the grand banquet that would run down the curtain on the most exhilarating tournament in rugby history. Quite a lot had happened that day, one way or another: the match, as tight as could be across the 80 minutes, had gone to extra time; the South Africans, supercharged both physically and emotionally, had spent the afternoon neutralising the New Zealanders' principal threat by gang-tackling Lomu into oblivion before Joel Stransky dropped a goal to win it for the hosts at the death. Above all, there had been the Mandela factor. By taking the field before kick-off clad in a Springbok jersey bearing Pienaar's No. 6, he sent the temperature of the occasion soaring clean off the gauge. You hardly needed a degree in political science to appreciate the symbolic importance of the new president's gesture.

Yet at the banquet, the rugby was barely mentioned. Not by the players who had been on the field that day; not by those of us who had looked down on events from the stands. Had it not been for a misjudged speech by the supremely graceless Louis Luyt that provoked a walk-out by the beaten and dejected All Blacks, at least one of whom was more than ready to thump the SARFU president on his way to the door, the only subject of discussion was the breakaway. Had a complete outsider with no interest in rugby happened to find his way into the dinner, I swear he would not have known that a famous match had been played that afternoon. In every corner, there were groups of players talking quietly about the immediate future rather than the immediate past: South Africans and New Zealanders, Englishmen and Frenchmen, Frenchmen and South Africans, New Zealanders and Englishmen. 'Are you in?' 'Yes. Are you?'

And in the background, the power brokers in the various national unions and their representatives on the International Rugby Board were experiencing what you might call an 'oh shit' moment. On the eve of the final, there had been an announcement of a major investment in southern-hemisphere rugby by Rupert Murdoch, whose rivalry with Packer, his fellow Australian, seemed to know no bounds. But with the authorities insisting that the new money was entirely consistent with regulations on amateurism and that the future shape of the game would be decided by them, and them alone, at a meeting later that summer, the breakaway plan still had solid support. I have no doubt that had Turnbull said to those of us still in South Africa, 'Here's proof of the money, here are your contracts and here's where you sign, so we'll all meet at this hotel tomorrow morning and get it done', a high proportion of those signatures would have been forthcoming. But he didn't say that, and the contracts never appeared. What he did say was: 'It's all on, but right now you need to go back home. We'll keep the dialogue open and we'll have it sorted within a month.'

To this day, I'm not sure whether he was playing for time because there were holes in the finances, or whether it was a tactical error on his part, or whether we had been led a merry dance right from the start. All I know is that we were fired up and ready to commit, and that we flew back to England feeling a little deflated.

I was on the phone to Will, our principal point of contact, and Brian pretty much every day throughout July. I wasn't aware of any discussions with the RFU: I certainly wasn't talking to Twickenham about this subject and I don't think

Will was either. I'm not sure the union would have engaged with us anyway. We were focused on getting the deal across the line and working out ways of bringing in players from outside the England squad to help make up the numbers. We intended to explain to them why and under what conditions the top end of the sport in our country was turning professional, and to ask them if they were willing to come with us. There was a risk of the union picking off the next strata as a means of cutting off the head of the snake – even an organisation renowned for moving at the speed of continental drift would have to do something, surely? – and we were fully aware that such a move would cause a major split in the domestic game. But we were determined to press ahead, and it was only when the trail went cold in late July that we smelled a rat. Suddenly, Will couldn't get hold of people in the southern hemisphere and the rumours began to circulate, slowly and painfully, that the top players on the other side of the world were U-turning towards their governing bodies and reaching new agreements on the basis of the Murdoch investment. There was a buy-off underway, and we weren't part of it.

On 27 August 1995, the IRB declared rugby union an 'open' sport. By which they meant professional; not that they could bring themselves to let the word pass their lips. What was more, it was 'open' with immediate effect: no planning, no preparation, no long-term strategy. There wasn't even a short-term strategy. The unions had saved their own necks by finding a way to reassert a semblance of control, but only because they had found themselves in panic mode and agreed to the very thing we players had been trying to move towards for years. Along with the likes of Will and Brian, I'd been half anticipating some acceptance of player payments since our trips to New Zealand in 1993 and South Africa a year later, but I don't think any of us foresaw a situation where the sport would be amateur at a minute to midnight on the 27th and pro at one second past midnight on the 28th. As for our first reaction . . . well, I can give you mine. I was sitting at home thinking: 'We've been badly let down by our friends in the south, who have accepted huge amounts of Murdoch's money while the Packer thing goes up in smoke. Me? I'll be on the tube to work tomorrow morning, as per usual. My £100,000 gig won't be happening. I'm still a surveyor, so I'd better get to the office on time and then head off to Wasps for training, because the new league season starts next week and, anyway, there's nothing else to do. Oh well. It was fun while it lasted.'

The dirty had been done on us, no question, but when all is said and done, those weeks before and during the World Cup in South Africa remain among the most extraordinary I've ever experienced. Leaving aside the Packer-related business, the tournament had taken on a life of its own right from the start and, when I think back to the final, I wonder whether it is indeed possible that some things in sport are written in the stars. I'm not naturally inclined towards astrological explanations for anything, let alone the result of a game of rugby – but the Boks must have felt some kind of force behind them when Andrew Mehrtens, the New Zealand outside-half, messed up a drop at goal just before the end of normal time, a kick he would definitely have nailed nine times out of ten and probably have landed 99 times out of 100. Had the situation been normal, the South African balloon would have been burst once and for all. But the situation was anything but, and the all-encompassing abnormality produced another of those moments in time, similar to the one that had happened to me against the Wallabies a couple of weeks previously and that Jonny Wilkinson would encounter eight years hence.

But then, there was a 'moment in time' feel about the whole competition, and as I flew home, I could not help asking myself if I'd dreamed at least some of it. We'd spent the back end of May and most of June up to our necks in a World Cup in a country deep in the throes of transition – a phenomenal period of change that will still be a subject of historical study a century and more from now. And as if that wasn't enough, a number of us had spent a lot of time pondering a move that would have been transformative both for ourselves and for the sport as a whole. I was pretty tired by the time we touched down at Heathrow, with good reason and, if I'm honest, I could have done without what followed. 'Where's the money?' 'Don't worry, it's coming.' 'Hang on a minute, this doesn't sound right.' 'It's all off.' 'That'll be that, then.' It was not an edifying few weeks.

And then, by way of putting a tin lid on it, the RFU responded to the 'open game' declaration by calling a one-year moratorium on professional rugby in England. Maybe the council members failed to understand what was happening. Maybe they were in a state of denial. Maybe they thought that if they ignored the southern hemisphere and its new money, it would simply go away. Whatever the reason, it was a monumentally bad decision – one that justified Will's now infamous pre-World Cup remark about age and flatulence. If there had been a table handy, I'd have banged my head on it . . . and kept banging until I'd knocked myself out.

While the big names in South Africa and the Antipodes were celebrating their new-found riches with fat cigars and vintage bubbly, we were left to contemplate 12 months of suspended animation. When I suggested to people at Twickenham that they should sign the current international players and those close to the England squad on central contracts without further ado, it fell on deaf ears. It is a little-known fact that Don Rutherford, modern-minded enough to recognise that the status quo would not last indefinitely, had drawn up a contract along those lines a year or so previously, and had a bunch of these documents, 35 or 40 of them, sitting in his bottom drawer. Which was where they stayed.

By choosing to fudge the issue, to kick the can so far down the road that my drop goal against the Wallabies was nothing more than a toe-poke by comparison, the senior figures at Twickenham left themselves exposed to forces that would weaken their authority and leave them at the mercy of a new breed of sporting entrepreneur. It was a critical failure at a critical juncture and the consequences of that failure are still unfolding. When, in years to come, some leading owner-investor or other tells the RFU that the club game doesn't need central money because it has plenty of its own, and that the Six Nations will either be played in five weeks flat or it will be played without a single Premiership player; and that, by the way, the England coach will have two days with his team before an international rather than two weeks; and that the top sides in Europe, or perhaps further afield, have negotiated a mega-deal with a major broadcaster and will be launching a super-duper world league sometime before supper – what will be said then of the fear-induced paralysis at the home of rugby? As a governing body, the union was guilty of failing to understand the nature of sporting enterprise and underestimating the ability of a bold group of free marketeers to get things done in minutes rather than decades. Alerted to the new opportunities created by the sanctioning of an 'open' game, the money men came marching over the hill. And this time, the money was unquestionably real and the contracts were more than ready to be signed.

3

WILD WEST IN THE NORTH-EAST

SIR JOHN HALL. Had anyone with a semblance of power at Twickenham ever heard of him? The smart money back in the summer of 1995 would have been placed on a 'no': I don't think it is stretching a point to suggest that few dyed-in-the-wool rugby folk would have spent so much as half an hour in the company of the miner's son from the Northumberland town of Ashington – birthplace of football's World Cup-winning Charlton brothers, no less – whose sporting interests did not include, as far as anyone knew, the mechanics of the scrum or the subtleties of the double miss-move in midfield. The RFU would get to know him soon enough, though. The declaration of the 'open game' had yet to percolate fully through the body politic of European rugby when Sir John, proud owner of Newcastle United FC, added something slightly less glamorous to his business portfolio: Newcastle Gosforth RFC. This was the most recent incarnation of the age-old Gosforth club, who had been double John Player Cup winners in the 1970s, when British and Irish

Test Lions as celebrated as Peter Dixon and Roger Uttley could be found in their forward pack, but had long since lost their place at the cutting edge of the domestic game in England.

There were a couple of things about Sir John that separated him from the overwhelming majority of the people running the sport from Twickenham at the time. He had ideas, for one thing. For another, he was entirely unafraid of backing his own judgement in the face of widespread opposition, and loved a fight almost as much as Brian Moore. He was a self-made man, in common with so many successful entrepreneurs: he had worked in the mining industry as a surveyor, scrapped his way up in property development and, following a study trip to the United States, launched his signature project – the construction of the first American-style out-of-town shopping centre in the United Kingdom, choosing as its location the waterlogged slag-heap site of an old power station in Gateshead. Many locals considered it the craziest idea ever, not least because there was no proper road access, and there came a point when Sir John almost ran out of money. But the Church Commissioners of England stepped in with the readies to complete the project, a major new bypass was constructed at just the right moment and Sir John's vision duly came to pass. Now, he was a major player in the business community – a sufficiently big hitter to have purchased Newcastle United and persuaded a figure as widely celebrated as Kevin Keegan to return as manager to St James's

Park, where he had spent a couple of seasons at the back end of his outstanding playing career.

Little or none of this had registered with me at the time. I was a man of the north myself, albeit a Yorkshireman rather than a Tynesider. I'd been educated at Barnard Castle School up in Teesdale and I'd played some of my early club rugby for Middlesbrough at a time when they were really quite good. But I'd long since set up camp down south, earning a crust in London and playing for Wasps, and events in Newcastle were a long way from the forefront of my mind when, one Tuesday night after work, I switched on the television and caught an item on Sky News about Sir John's purchase of Newcastle Gosforth and the wiping out of the debt – substantial for the club, if not for the new owner – that had accumulated over recent years. I was intrigued, not only by the amusing footage of Sir John and Godfrey Clark, the Newcastle Gosforth president, flicking a rugby ball backwards and forwards as they ran across the St James's Park pitch (not very quickly, it has to be said; still less with the panache of a pair of French centres), but also by the realisation that this had happened so quickly. The International Rugby Board's document ushering in the age of professionalism had been tabled only a week previously, and here was another bolt from the blue.

I remember thinking: 'I've been let down by the southern hemisphere boys, the RFU have declared their moratorium, there are central contracts in Don Rutherford's bottom drawer that may never see the light of day. Wasps have no money, there's no television coverage of English rugby worth mentioning, I'm back in the world of nine-to-five . . . And here's this bloody big story. Here's a bloke who knows how to cut a business deal, seems really straight and is passionate about the north-east.' And while I was watching, something happened. It was bizarre, a bit weird. It wasn't really a premonition, but it was one of those moments when I thought: where could this lead? It was out of left field, it was interesting, it was happening back in my old neck of the woods and my mind was racing.

The very next day, as I returned to the office after a lunch meeting, my PA said: 'I've left a note on your desk. Someone called Freddy Shepherd from Newcastle United has rung you and wants you to get back to him.'

It put me in a cold sweat. I could see the yellow sticky note by the phone, with the number written down. Even now, decades on, I can remember nearly every digit. My heart was beating hard as I punched in the area code: after the strange feeling of the night before, I was just a little spooked. But Freddy, a man with some family money behind him but largely self-made like Sir John, turned out to be a down-to-earth type who liked to put people at their ease. 'We thought it was a good idea to buy this rugby club because we heard the game was going pro, but we're not really sure what to do with it now we've got it and have no clear focus on how to run it,' he told me. 'We'd like to talk to you about all this – to find out what you think about what we've done, whether you'd give us some thoughts about where you think rugby is going . . . whether you might even be interested in playing for us.' When I expressed some interest, he immediately invited me to a meeting. 'Newcastle United are playing Southampton away this weekend, so we can get together in London easily enough,' he said. 'What are you doing on Friday night?'

When I put down the phone, my thoughts were all over the place. Newcastle were playing their rugby in Division Two of the English league, having been relegated from the top flight after a really horrible campaign the previous season, and were suffering their share of problems even down there among the second-class citizens. Wasps, on the other hand, had reached the cup final only a few short weeks previously, had finished third in the first-division table after a fierce scrap for the title with Leicester and Bath, and had some of the best players in the land on their books. There again, the weeks since the end of the World Cup had hardly been the best of my life. I'd gone from being within touching distance of a £100,000-a-year contract to contemplating a year-long RFU moratorium that amounted to diddly-squat.

After my conversation with Freddy, I definitely felt a pull. Was I simply on the rebound? Maybe. But I was 32, I'd played at three World Cups, toured twice with the Lions, won three Grand Slams and made 70 appearances for my country. I felt I had a bit left in me as a player but, even without the call from Tyneside, there would have been decisions that needed taking. Should I retire from international rugby on the basis that it's generally better to jump than be pushed? Was it time to pack in the game altogether and throw myself into my career as a surveyor? Did I really want to be a surveyor anyway? Was I really cut out for it? These were important questions, and they needed answering. The Newcastle business might have put my issues into a different context, but that did not make those issues any less fundamental.

We met, as arranged, in a private room at a restaurant in Victoria, where Freddy was joined by Sir John's son, Douglas. During dinner, they talked with infectious enthusiasm about their idea of a 'sporting club' venture that would give the good people of Newcastle the range of attractions famously enjoyed by the citizens of Barcelona. They said Sir John saw the rugby side as a perfect fit with his interests in football and two other sports that were very much in his sights, ice hockey and basketball, and made it abundantly clear he was determined to bring the plan to fruition. Sir John had deep admiration for the Barcelona model, which amounted to nothing less than an expression of Catalan power and individuality through sport, and was completely driven in his desire to establish something similar in the north-east. I sometimes think that he'd have voted for Geordie independence, if such a thing had ever been on offer. The People's Republic of Tyneside? He'd have been the first on the barricades, probably waving a flag of his own design.

The conversation at the dinner fed into the general sense of impatience with the world I was feeling at that moment.

The Packer proposal had energised me: it had raised the possibility of finally giving the establishment a two-fingered salute after all the lack of foresight, the lack of recognition, the undervaluing of the players. That particular opportunity was no longer on the table, and it was depressing to think that more and more would be asked of those in the England squad while the union continued making more and more money. What a great business model they had, selling out Twickenham while spending next to nothing on the entertainers! All things considered, I was definitely in the mood for something different. And by the time I left that restaurant in Victoria, I was hooked.

Hooked by the offer of a job with no defined role; captivated by the prospect of working on a blank canvas. It was entirely characteristic of my dining companions and their entrepreneurial spirit that when asked exactly what they might expect of me if I made the move to Newcastle, they replied: 'We'll work it out when you get there. In fact, *you'll* work it out when you get there. If you want to play, fine. If you want to run the place, fine. If you want to play *and* run the place, that'll be fine too.' It was virgin territory for all of us: we were talking about a newly professionalised rugby club with no professional structures, having been an amateur club with antiquated structures only a few days previously. If I didn't fully understand the parameters within which I'd be working, neither did they.

I was excited, to say the least. I was not exactly a born

militant and had never been completely comfortable with my occasional flirtation with industrial action, but there was undeniably something thrilling about this particular act of resistance to officialdom – about the fact that there were folk at Twickenham already kicking up merry hell about the advent of professionalism, and would kick a lot harder when this Newcastle thing was up and running. These were the people who could have taken control of rugby in the open era, who had been given a chance to shape the sport, to take it down a different path than the one it was about to embark upon. Would rugby union be in a better place now, had they proved themselves up to the task? We'll never know. What we do know is that the old guard missed their moment.

I had pretty much decided, there and then in the restaurant, that I wasn't going to miss mine. I had a friend, Geoffrey Hamilton-Fairley, who was a businessman – a bit of an entrepreneur himself, as it happened. We'd met a few years earlier, when he was acting as an agent for the Falklands War hero Simon Weston. While we hadn't done much work together, I felt I knew him well enough to phone him and say: 'We have something that needs some attention here. Can you deal with these football people on the contract side of things while I get my head around running a rugby club?' Within a few days, there was a five-year deal on the table, on financial terms I considered perfectly reasonable. And that was that. I was off.

Not that these things are ever entirely easy. My employers, DTZ, had never been anything other than completely supportive: in fact, they had gone a very long way beyond the call of duty in putting up with my all-too-regular absences on rugby business. It was not a great feeling, walking into the office and saying: 'I'm really sorry. I've been here ten years, during which time I've been away a hell of a lot. You've paid me the whole time, always guaranteed that I'd have a job to come back to and generally been fantastic. Now that I'm in my 30s, you have every right to assume that I'll be giving up this rugby lark and concentrating full-time on my career. But that's not what's happening, I'm afraid. I've had an offer I can't refuse for lots of reasons and they want me to start tomorrow, so I'm handing in my notice.' I've experienced more comfortable days, I must admit.

And then there was Wasps. My club for the last eight seasons, with whom I'd won a league title and had my fair share of excitement, would also need to know what was happening. In this respect, the situation was more than a little complicated. For one thing, I was now on the Newcastle payroll, even though players in England were still not allowed to earn money from rugby. Not from actually playing, at least. So my title on Tyneside was 'director of rugby development' – a title I came up with myself. What did it mean? To this day, I'm not quite sure. Was it legitimate, under the RFU regulations then in place? I'm not sure about that, either. But as this was the Wild West, with everyone scrambling around trying to fill the vacuum created by the great cop-out in the Twickenham council chamber, the obvious answer on being challenged would have been: 'Whether it is or it isn't within the rules, what are you going to do about it?'

The added difficulty was the 120-day stand-down period imposed on all players who switched clubs in those times. Knowing that I couldn't turn out for Newcastle, the Wasps hierarchy, fronted by the ever-dedicated head coach Rob Smith, decided they were happy to keep picking me. Which was fine, except that I had a new squad to build . . . and there were a number of Wasps teammates who ticked all the boxes in terms of the quality for which I was searching.

There were some useful players already in the existing Newcastle side: the scrum-half Steve Douglas, the prop Paul Van-Zandvliet, the unusually substantial lock Richard Metcalfe, the aggressive flanker Richard Arnold. They had the wherewithal to handle the pressures of rugby in the Premiership, which is where we intended to be within two seasons. But many of the others, honest-to-goodness as they were, did not quite meet the requirements. There was a bit of money around from the moment I arrived, and I was able to push a few bob their way by reinventing them as 'rugby development officers', but virtually all of them were committed to their proper jobs in teaching or accountancy or whatever. I had to find myself a reliable backbone around which I could construct a team capable of securing promotion sooner rather than later, and it just so happened that the people I felt I needed were on the books at Wasps: people like Dean Ryan and Steve Bates and the Lions Test prop Nick Popplewell.

These were my first recruits, along with the talented centre Graham Childs, whose capture was mercifully controversy-free because he was heading to the north-east anyway, having accepted a job in Sunderland with Nike.

Ryan and Bates were my initial targets, not least because they were natural coach material. I was heading up the thing, sure, but my coaching experience was limited at best. I remember approaching Dean during a Wasps training session: it was like something out of *All the President's Men* as I said, 'Meet me in the car park when you're done.' I asked him if he was interested in a move to Newcastle; he asked me what was on offer. I told him I could bring him a three-year contract on Thursday night. 'I can probably go to £50,000 a year,' I said. 'Will that do?' Dean decided it would do nicely.

It was blindingly obvious that Wasps would soon lose patience with the situation, yet when Rob Smith summoned me for a face-to-face meeting and told me I would not be playing for the club again, there was no bitterness, no air of animosity. Rather, it was far more civilised than I had expected, or had any right to have hoped. Do I believe I betrayed Wasps by luring away some of their best talent? Did I feel any sense of guilt, of shame? The honest answer is 'no'. The sport had changed overnight, I had a job to do. I'd made a big call to take it on and I had no intention of messing it up. My priority was the building of a Premiership-standard rugby club, pretty much from scratch. End of. And besides, it wasn't me who made the decision to throw the game open and leave it exposed to a new set of wholly unpredictable forces. There were people at the top end of rugby administration who had sat on their arses for years, with their eyes and ears tight shut. If no one was in control any more, they were the ones to blame. Those of us involved in the Newcastle project were simply reacting to the facts on the ground.

Those facts were not at all palatable as far as those on the arch-traditionalist, increasingly reactionary wing of the RFU were concerned, and it was not long before the coverage in the newspapers was dominated by rugby politics. There were times when it seemed the battles in the boardroom and the conflicts in committee were forcing the fixtures on the field off the sports pages altogether. Some prominent Twickenham figures – the likes of Fran Cotton, that great front-row forward whose exploits with England and the Lions had become the stuff of legend – argued that the game faced an existential threat from a small group of business-suited money men who were hell-bent on seizing control of the entire union code from the established authorities. Fears were expressed that international rugby in general, and England's role at Test level in particular, was being placed at risk by a bunch of bank accounts in human form who were 'not union people' and therefore had no attachment to the game beyond its profit-turning potential. There were high-level meetings every day of the week, often preceded by meetings about meetings and followed by meetings to discuss the implications of the meeting just concluded. It went on for what seemed like an eternity and, as is often the case with disputes of this kind, the rival positions became ever more entrenched.

I'm happy to say that to a large extent the politics passed me by, at this stage at least. I had quite enough on my plate without having the affairs of state eating into my time, and anyway, the owners and major investors – Sir John, Nigel Wray at Saracens, Tom Walkinshaw at Gloucester and Andrew Brownsword at Bath among others – quickly recognised the importance of maximising their strength and influence by banding together as a group, and it was not long before they were speaking with one voice, as wealthy individuals tend to do when the situation demands. I was more than happy to give Sir John advice when he asked for it but, looking back, it was more a phoney war than a real one. The Twickenham traditionalists were always playing catch-up: having left the stable door open and seen the horse disappear over the distant horizon, they had no workable plan – no plan of any kind, it seemed to me – to get it back in harness. It was all very well them criticising the club owners at every turn, but there was something futile about their response. They were raging against the dying of the light.

As the squabbling continued, I proceeded with the day job: piecing together a team strong enough to get us out of the second tier and into the big time over the course of the 1996–97 campaign. This was no small challenge – in my first season, a real 'suck-it-and-see' job if ever there was one, we'd finished eighth out of ten, on the same number of points as

Nottingham and Bedford, the clubs below us, but as I settled in and developed a clear idea of exactly who I wanted to attract to Kingston Park, things began to move in the right direction. Tony Underwood, the England wing, joined us from Leicester, as did a couple of high-class Scots in Gary Armstrong, the international scrum-half, and Doddie Weir, the future Lions lock, both of whom had been playing their club rugby just up the road in the Borders.

Therein resides an amusing tale. Sir John was big into press conferences – he loved the idea of us blowing our own trumpet, as loud and long as possible, whenever a significant signing was completed. So when the Underwood move was nailed down, we gave it the big production. (I almost did the double by recruiting Tony's brother and record-breaking England try-scorer Rory, with whom I went back a very long way, but my old Barnard Castle schoolmate was still heavily involved in his pilot's work with the RAF and didn't feel he could commit himself to full-time professional sport.) I was also closing in on Gary and Doddie at this point, and thought I might clinch the deal by showing them round St James's Park, where we sometimes trained, just as Tony was being unveiled to the media. Sure enough, they rocked up to the stadium to be confronted by television cameras and reporters' notebooks. 'You bastard,' said Doddie, politely. 'We're not there on the money yet.' Which was no more than the truth, for he was holding out for an additional £5,000. The deal was not clinched until I arrived at the stadium for my regular

Thursday morning update meeting with Freddy and one or two others. Asked how the negotiations with the two Scots were coming along, I said: 'We're almost there, but I've reached a sticking point with Doddie. We're five grand apart.'

'Five grand a week?' asked Freddy.

I blinked. 'Not quite,' I replied. 'Five grand a year.'

Freddy looked at me as though I'd lost the plot. 'Five grand a year? A *year*? Just give it to him, for Christ's sake.'

It was a reminder, if I ever needed it, that the only thing English rugby finance has in common with the football version is the use of the same currency.

There was another striking expression of the go-get-'em mentality from the board when, a season or so later in the Newcastle story, I drew a deep breath and began my pursuit of Va'aiga Tuigamala, who, with Jonah Lomu seemingly in decline because of the serious health problems that would ultimately wreck his career and end his life at the depressingly early age of 40, was just about the biggest name in world rugby, be it union or league. Born in Samoa and fiercely proud of his island heritage, Inga had been a mighty figure in both codes since first surfacing as an ultra-powerful, prototype Lomu-style wing with the All Blacks in the late 1980s. He switched to league after the Lions tour of New Zealand in 1993, joining Wigan on the kind of money that made our commercial return as England players look like the pocket money it was, and he was still playing for the club when the union game went open and we saw the first signs of a reversal of the old drift from the 15-man game to the 13-a-side variety. Jason Robinson and Henry Paul, two of Tuigamala's colleagues at Central Park, materialised at Bath on short-term contracts; another big Wigan name, Gary Connolly, headed south to Harlequins. Inga had reintroduced himself to union once the shackles of amateurism were removed – he played a few games for Wasps, of all people – and news soon reached Tyneside that my old club were interested in bringing him on board full-time.

We were already into one outstanding Samoan international who could boast All Black credentials into the bargain: Pat Lam, the inspirational back-row forward who had made a heavy impact at consecutive World Cups, was at the top of our target list and happened to be on tour in England. I sent Steve Bates and Dean Ryan down to Oxford, where Samoa were playing a midweek game, with a view to opening a dialogue with Pat. When they returned, they left me with the distinct impression that Inga might also be open to a move in our direction. The Samoans were playing Ireland in Dublin the following weekend, so I hopped on a plane and just happened to book myself a room in their hotel. By some strange coincidence, I also found myself sharing a lift from the lobby with . . . Pat and Inga. 'Fancy a chat?' I asked, my face a picture of innocence.

The upshot from Inga's perspective was that, while he was keen to join us, there was a four-year contract with Wigan standing in the way. Given the money he was commanding, this was problematic. But as luck would have it, there was a full-force gale of extravagance blowing through Newcastle at that moment. A few weeks previously, the football club had splashed out a record £15 million on Alan Shearer, bringing the Euro '96 hero back to his home town from Blackburn Rovers, with whom he had won the Premiership title. Sweeter still, they had beaten Manchester United to his services, and there was ambitious talk from the St James's Park boardroom about the club becoming a major force in the European game. The city was alive with sporting passion: 10,000 supporters turned up just to catch a glimpse of Shearer on the day he put pen to paper. I remember attending a black-tie function at Sir John's country estate on Teeside and talking to Douglas Hall, the chairman's son and a prominent figure in the sporting side of the business. 'We've bought ourselves the most expensive footballer in the world,' he said to me. 'Now, we want the most expensive rugby player.'

I told him I knew just the man: Inga. 'He'll be plenty expensive enough,' I confirmed. 'I think we can get him out of Wigan, but it will be at a price.'

'Go get him,' said Douglas.

'Are you serious?'

'Yep. Just tell me what it will cost.'

The cost was a £750,000 transfer fee, together with a very handsome salary wrapped up in a five-and-a-half-year deal. At the same time, Richmond paid Wigan around £400,000 for the Welsh No. 8 Scott Quinnell. Two players sold back to union; more than a million in the Wigan account. Tasty business indeed. Did the men who declared an 'open' game in August 1995 imagine such sums changing hands as early as 1996? Somehow, I doubt it.

Over the course of the 1996–97 season, there was solid evidence that we were on a good road. Richmond were our most potent rivals, armed not with one Quinnell but two – Scott's brother Craig had also made the move to south-west London – and a number of other high-profile reinforcements, from my old England colleagues Brian Moore and Ben Clarke up front, to players as classy as Adrian Davies, Allan Bateman and Simon Mason behind the scrum. Meanwhile, we had attracted additional international talent in John Bentley and Garath Archer at wing and lock respectively. We met early in the piece at the Athletic Ground, which Richmond shared with London Scottish, and the place was packed: 7,000 spectators sardining themselves in for a game that would, not so many months previously, have pulled in 700 if it was lucky. The outcome? A draw, 20-apiece. And we would remain locked together for the duration. Richmond pipped us to the title by a point, largely because we lost to a Coventry side that was more than half decent in those days, but with two going up, it was not the end of the world. We certainly felt we had outgrown the second division: there is only so much satisfaction you can take from scoring shedloads of points against weak opposition. One afternoon, we put 24 tries past Rugby in a 156-5 victory. I kicked 18 conversions. Enough said.

It was a sign of our strength that when the British and Irish Lions left for their tour of South Africa in the late spring of 1997, which was their first visit to that country for the best part of two decades, there were five Newcastle players on the plane: the full-back Tim Stimpson; Bentley, Underwood and the versatile Alan Tait in the three-quarters; Doddie among the engine-roomers. As Richmond contributed a tourist or two themselves, with Bateman and Scott Quinnell both making the cut, a fifth of the original squad were drawn from the second tier of the English game. Could such a thing conceivably happen again? I won't be holding my breath.

At that particular point in the history of rugby union in these islands, anything and everything seemed possible. Those of us in the vanguard of the revolution were making it up as we went along and, while the building of a new club for a new age was exhausting work, it was also intoxicating. When we celebrated our progression to the freshly minted Premiership at the end of the season, we felt we were on a runaway journey to somewhere special. Given that virtually the whole squad had spent years labouring under the restrictions of the amateur ethos, I think we could be forgiven for thinking we had finally reached the land of milk and honey.

With the committee-room arguments still in full flow, we were not winning quite as many popularity contests with the Twickenham classes as we were winning matches. But were we really insurrectionists? Were we really nothing more than a bunch of slashers and burners who were prepared to destroy the entire fabric of English rugby in exchange for financial reward? It was a well-rehearsed argument at the time, and I guess there are still some members of the RFU's county set who look back on that period with a deep-seated loathing of everything we stood for and everything we did. All I can say now is that the Newcastle project was nowhere near as ruthlessly acquisitive or cold-heartedly destructive as its critics made out. The declaration of the 'open' game created chaos. Not organised chaos, for there was nothing remotely organised about it. Just chaos, pure and simple. Sir John bought himself a stake in rugby primarily because he thought there was some fun to be had, and from his point of view he was right: in the four years he spent with us, he derived immense pleasure from the camaraderie, the travelling and, yes, the success. From the players' perspective, they were simply grateful to be paid decent money for doing something they loved, rather than being forced to do it for free. They were all keen to secure the best deals they could, but precious few, if any, fell prey to greed. Doddie and his five grand? We still laugh about it whenever we meet up. It was hardly an example of casino capitalism.

Shortly before the announcement of the Lions squad, I found myself back in the international fold. After 18 months or so of shocks and surprises and jaw-dropping events, this one really caught me off balance. I received a phone call from Mike Slemen, a Grand Slam-winning England wing in 1980, who was now on Jack Rowell's national coaching team. There were next to no outside-halves left standing, he told me – everyone except Mike Catt was injured. Would I be prepared to sit on the bench against Wales in Cardiff, the final game of that season's Five Nations? I'd retired from the international scene soon after throwing in my lot with Newcastle: to be precise, I'd issued a statement on the day of my meeting with Rob Smith at Wasps, for the simple reason that I'd soon be spending every waking hour at one remove from the top flight of the domestic game. Mike's call was completely unexpected but, following a quick scratch of the head, I agreed.

After joining up with the squad in Marlow (the captaincy had passed to the Bath centre Phil de Glanville by this time, but Will Carling was still in the side, alongside such familiar faces as Jason Leonard, Martin Johnson and Tim Rodber), we travelled west on the well-worn road to the Arms Park. It was a strange feeling, being back in those parts amid the swirl and snarl of match day, but we won well, 34-13, and I was given a seven-minute cameo at the end. I remember precisely nothing about it, apart from the fact that the occasion of my 71st appearance in the white shirt felt more bewildering than the previous 70 put together. It would also be Will's final cap. Jonathan Davies, the brilliant Welsh No. 10 with whom I'd tangled in the past, not always successfully, also bowed out that day. For all that, I do not remember it as an especially emotional occasion. I had other items on my agenda.

4
THE JONNY FACTOR

WE MEASURED OUT our lives in everlasting coach journeys. Newcastle was a long way from almost everywhere in terms of top-flight club rugby. Two-thirds of our rivals were either in London, the West Country or somewhere along the M4; even Sale, our fellow northerners, were well over 100 miles away on the far side of the Pennines, so there was abundant road time for us to get to know each other. Our swankier 'sporting club' colleagues at Newcastle United had it a little easier, of course: they generally travelled to away games by plane, with the team up front and the guests and sponsors down at the back. Not even the board members were given a seat within hollering range of the pilot when the Magpies were in the air: the Keegans and Shearers were virtually on the flight deck; the men underwriting the operation were among the baggage. Truly, football and rugby existed in different universes, and the parallels were not obvious. For all the progress made on what might be called workers' rights in the union game over the last 20 years, they are no more obvious now.

Two return trips stick in my mind. Actually, that's not quite accurate. The memory of the first of those treks, home from Bath on the opening day of the 1997–98 season, remains as clear as day: by the time we found our way back to Tyneside after beating the team who had completely dominated the English domestic scene since the early 1980s, we understood far more about ourselves – the physical edge we could bring to our rugby, the depth of our togetherness, what made us tick as a group – than we had 48 hours previously. The second, following the last title-clinching match of the campaign at Harlequins, has always been just a little hazy, thanks to a heady mix of jubilation and beer. A hell of a lot happened between those points in time, including the emergence of a certain Jonathan Peter Wilkinson. One way or another, he would have quite an impact on life at Kingston Park.

But Jonny was barely in the back of my mind, let alone the forefront, as the team boarded the charabanc in late August for the long drive to the Recreation Ground, which was just about as far away from Tyneside as it got in those days. At that point, he was fresh out of school and some months away from forcing his way into our starting midfield, partly because I was still doing a turn at outside-half and acting as principal goal-kicker; partly because Inga Tuigamala, one of the biggest names in world rugby as well as one of the biggest backs, and Alan Tait, a major contributor to the British and Irish Lions' unexpected series victory over the Springboks earlier that summer, were obvious first picks at centre; and partly because he was just a kid, albeit an unusually mature one as far as his approach to the game was concerned. I had been clear right from the start of the Newcastle project that if we were to justify our place in the top flight and have even a remote chance of fulfilling Sir John Hall's ambitions for us, experience would be the key ingredient. But it was difficult, as a newly promoted side, to settle on a precise level of expectation: I knew we were strong enough to stay up – I'd signed enough tough-minded, highly motivated individuals to be confident of that much; but I don't think any of us had a clear idea of where the next nine months would take us. This was no time to be fast-tracking infants.

The prospect of playing Bath first up was quite enough to be going on with. They had been undisputed rulers of the roost as the amateur era stumbled to its inevitable conclusion in a fog of administrative confusion and committee-room inertia, and while they had fallen off the professional pace by failing to attract meaningful investment until very late in the day (at one point, it seemed possible that their entire first-team squad would leave The Rec in search of secure employment), they had stabilised themselves when the local businessman Andrew Brownsword had taken up a majority shareholding and had managed to keep hold of international-class players as accomplished as Matt Perry, Phil de Glanville, Mike Catt, Victor Ubogu, Nigel Redman, Richard Webster and Eric Peters. All of those names were on the team sheet the day we turned up on the banks of the River Avon, so to leave town with a 20-13 victory was quite something. The game had been broadcast live on television and generated quite a fuss, thanks to a fairly significant punch-up for which Dean Ryan carried the disciplinary can. There was nowhere near as much controversy when Bath claimed retribution later in the season by going after Dean in a big way at the start of the return match and leaving him in a really bad state, but that's rugby. Or rather, it was rugby then.

Sir John loved it, of course: there was nothing on earth he enjoyed more than winning. He was on the bus with us – if he didn't truly understand the fine detail of the game, he revelled in the camaraderie that lay at the heart of it, and particularly enjoyed playing cards in the back seat with Doddie Weir and Gary Armstrong, even though he routinely lost. (It might not seem the brightest idea for a mere player to make a raid on the owner's wallet, but Doddie was adept at getting away with most things, thanks to the expression of natural innocence that never appeared to leave his face; and anyway, he took the view that his employer could afford it.) As we made our way out of The Rec car park and into the narrow streets of the Georgian city, Sir John stopped the coach outside a branch of Marks and Spencer, dived headlong into the store and emerged with enough booze to last us the six hours or so back to the north-east. If our subsequent trips to France for European Challenge Cup matches in Perpignan, Biarritz and Agen would be equally competitively challenging, they were also every bit as entertaining on the social front. But we lost a couple of those matches by a small handful of points and it was in defeat that we saw the flipside of Sir John's passionate side.

If he loved winning, he hated losing. Hated it with a capital 'H'. More than that, he seemed incapable of accepting it. We didn't let too many games slip that season, thank heaven, but when we did, he'd try to find a way into our inner sanctum and make a point or two to the players. Losing dressing rooms are not the happiest of places at the best of times, so the last thing we needed was to feel the sharp edge of the owner's tongue. Even worse was the possibility that he'd approach the wrong bloke – Dean, let's say, or Nick Popplewell, or George Graham – and jab a finger in his chest at the wrong moment. Sir John might have been capable of riding a punch from one of his fellow investors, but would he stand up to a bunch of fives from a hard-nut forward in a bad mood? I had dark visions of someone chinning him, of a newspaper front-page splash under the headline: 'Sir John Hall knocked unconscious by his own prop.'

So I came up with a plan. I asked Steve Black, our minder and spiritual father figure as well as our strength and conditioning coach, to stand outside the dressing room after a defeat and keep Sir John out at all costs. 'Put your arm round him, stick him over your shoulder, walk him round the pitch for 20 minutes . . . do whatever you have to do, but don't let him near us until everyone's calm,' I'd tell him. Which was what happened. Blackie, the ultimate people person as well as a bear of a man, was brilliant at it.

In a way, you can understand it when a successful businessman reacts badly to defeat: even club owners blessed with a highly developed understanding of the game like Dave Thompson, who would succeed Sir John as the owner-chairman at Newcastle and spare us the agony of appearing in the bankruptcy courts, find it tough to keep their passion in check. When a sports club is costing you a packet (and at that time, Sir John was pumping in £180,000-plus a month), it cannot be easy to detach yourself from events on the field, to push your emotions to one side and see things in the cold light of reason. The pressure of the moment gets to the investors, just as it gets to players and coaches and supporters, but it's slightly different for them for obvious reasons. I've been in the company of people with heavy financial interests in a range of sports, including football, where the sums at stake are astronomical. They live every second, they throw every pass and kick every ball – at times, you can almost see the steam coming out of their ears. The most rational ones, like Nigel Wray at Saracens, handle it well, but as Nigel himself has admitted, it took him the best part of 15 years to reach a place where he could take the rough with the smooth. And how much did it cost him in pounds, shillings and pence over that period? I shudder to think.

Happily for those of us involved at Newcastle, the eruptions were few and far between in that first amazing top-tier season. The confidence we drew from beating a Bath side who were good enough to finish the campaign as European champions set us on a roll and, because of it, together with the fractured nature of the Premiership fixture list, we did not lose a league game until the middle of March, when we came unstuck on a trip to our old friends at Richmond. Even though Saracens had emerged as our most serious rivals at the business end of the table and were pushing us hard, the defeat in south-west London seemed, on the face of it, to be far more of an inconvenience than a calamity. We still had every right to feel good about ourselves, having won at Leicester shortly after Christmas, a victory made all the more blissful by the fact that the Tigers had armed themselves to the back teeth with such natural-born winners as Joel Stransky, Waisale Serevi, Martin Johnson, Dean Richards and Neil Back, and could not even begin to conceal their disgust at losing to us. We had also survived an awkward trip to Northampton, where even the best sides could slip up all too easily.

The endless motorway mileage had played its part in tightening us as a group, and we'd developed a balanced style of rugby that forced our opponents to answer all manner of very difficult questions. Dean and Pat Lam were really clicking in the back row; our scrummaging was tidy enough; Gary was working the fringes in the way he did for Scotland; and we had more than enough strike power in the backs, with Tim Stimpson, Tony Underwood and John Bentley feeding off Inga and Alan in the centre roles. We didn't overdo the flashy stuff; instead, we carried hard and played the corners for all we were worth. If, every once in a while, our pack found it difficult to establish supremacy, we could rely on Inga to put us on the front foot. Looking back, we must have been a serious pain to play against. Yet as we headed into the last two months of the programme, I felt we were slipping off our standards – that there was a glitch in the system. I decided to break up the established centre pairing and promote Jonny to the starting team for the run-in. It was then that the Wilkinson story began to unfold.

He was still a newcomer: just a few months previously, he'd been in an A-level classroom at the other end of the country. (As far as we were concerned, everything that mattered in rugby seemed to happen at the other end of the country.) But we'd known about him for a good while, even so. Steve Bates had made the introduction, while we were playing together at Wasps. Every bit as masterful a teacher as he was a scrum-half, Steve had left Radley College in Oxfordshire to take up a position on the staff at Lord Wandsworth College on the Hampshire–Surrey border, and quickly spotted some rugby talent in the form of a 14-year-old outside-half who could kick the ball miles. 'I've got this kid at the school,' he said to me at training one evening. 'He's very talented. More than that, though, he has the most extraordinary work ethic. I don't think I've ever seen a youngster who puts so much into his rugby.' I was more than happy to take Steve at his word and, being senior players at Wasps, we naturally hatched a plot to bring the Boy Wonder to the club the moment the time was right. Unbeknownst to me at the time, Jonny had already hatched a plot of his own. A plot rather grander than the one we had concocted.

By the age of seven, he had decided on international rugby as his destination in life. His mother Philippa would drive her younger son, still in the early stages of his primary education, to the local rugby club in Farnham, and sit in the car until he'd finished his goal-kicking practice. Sometimes she would wait an hour or so; sometimes she would still be there as the evening light faded into the night sky. When he was a little older, she would leave him to it and return when she thought there was a 50-50 chance of him being prepared to call it a day. On most occasions, she'd find herself waiting once again. Right from the start, the extra mile was not nearly far enough for Jonny. He preferred the extra marathon, the extra circumnavigation of the globe. When, after we had left Wasps, the moment came for Steve and myself to point him towards Newcastle (and I should say at this juncture that we feared we might lose him, Surrey to Tyneside being a more challenging proposition for a southern boy than Surrey to London), that fanatical, almost maniacal approach to training and preparation was so ingrained in him that we would not have been able to lighten his self-imposed load even had we felt it necessary.

It was a new one on me, this obsessive streak. There would be occasions, a little later in his career, when I was tempted to wonder if he was driving himself much too hard – whether, in his case, the fine distinction between genius and madness was becoming dangerously blurred. But when he first arrived, the intensity of his attitude seemed like a 24-carat positive, and anyway, who of us can state with complete certainty where another individual should draw his parameters? I considered myself to be as driven as most. I was ferociously competitive as a youngster (I still am), and when it came to the work-ethic side of things, I probably put in as much as anyone I knew, and a fair bit more than most. But while I had the sense that I wanted to make the most of whatever I possessed, I had no clear idea of what that actually meant. I certainly don't remember telling myself at the age of seven – or even at the age of 17 – that there was a place in the England team with my name on it provided I put in the requisite amount of effort. A spot on the Durham Under-18s team sheet was just about the limit of my ambition during my A-level years and, while I found a way into that side, I also managed to find my way out of it again, dropped for Richard Cramb, who would go on to play a handful of games for Scotland. By the time Jonny was 16, he was already eight or so years into his long-term project, playing age-group rugby for his country and dropping a goal to beat the Wales schoolboy side . . . in Wales. My first England representative rugby was at Under-23 level. Spot the difference.

Having mapped out his journey north between us, it was down to Steve and myself to persuade Jonny's parents that Newcastle was situated some way south of the Arctic Circle and was not, therefore, wholly beyond the bounds of civilisation. Fortunately, Steve had built a good relationship with the Wilkinson family, and when we explained to Philippa and her husband Phil that our team-building plans were well advanced, that our ideas about becoming a major force in English rugby were something more than pie in the sky, and that Jonny would be well looked after from the player development point of view, they agreed to the move. During the Easter holidays of his final school year, he joined us at Kingston Park for a spot of training. We chucked him straight out there among the big boys and he looked entirely comfortable. I remember Pat Lam, who had come across his fair share of talented teenagers in both New Zealand and Samoa, saying to me: 'Hey, who's the kid?'

'Just someone we're taking a look at,' I replied.

'Well, there's no need to look for much longer,' Pat said. 'His skills are unbelievable.'

That summer, Jonny toured Australia with England Under-18s. It was a half-decent bunch of age-groupers to say the least: among the headliners were Iain Balshaw, Mike Tindall, Andrew Sheridan, David Flatman, Lee Mears, Steve Borthwick, Alex Sanderson and . . . an outside-half by the name of James Lofthouse, who claimed the No. 10 shirt for the outstanding victory over the junior Wallabies at North Sydney Oval. Bumped into the No. 12 position, Jonny would have been less than amused at this selectorial assault on his sense of perfectionism, but when he finally joined us on a one-year contract that would earn him the princely sum of £12,000 (a more than generous offer, we felt, for someone fresh out of the playground), there was no hint of lingering frustration. We put him up in a house in Newcastle's West End, which he shared with Chris Simpson-Daniel, brother of the future England wing James and a highly promising half-back in his own right. It was not the most salubrious part of town by a very long chalk; in the fullness of time, when Jonny found himself in a more sought-after property on the golf course at Slaley Hall, he must have felt relieved to be more at risk from a mistimed three-iron than a stray brick. Still, we all have to start somewhere.

His parents had decided to stay in Farnham for the time being – Phil would not sell his insurance business and head north for a while yet, and Jonny's brother Mark was not yet on our books, although he would join the club soon enough. But we need not have worried about him struggling to settle in the absence of home comforts and familiar faces: Steve Black saw to that. Blackie was a real find for all of us, but in Jonny's case he was a crucial figure, central to pretty much everything that would happen over the coming years. He was a formidable sort in all manner of ways. A hard case who had spent his later teenage years working the doors in and around the Newcastle nightclub scene, and who had encountered his fair share of trouble as a consequence, he was also blessed with enormous sensitivity – a gift that underpinned his remarkable powers as a motivator, a sounding board, an all-round positive presence. Blackie was a born optimist who knew how to bolster the confidence of those in his orbit. Jonny's sense of optimism was not quite so pronounced, to put it mildly, and while it seems odd to say so in light of what he would go on to achieve, self-belief was not his long suit. All things considered, Blackie was the antidote to the world according to Jonny – an oasis in a desert of torment.

He was already caught up with Newcastle when I arrived from Wasps, putting the existing players through their paces on Tuesday and Thursday evenings and accompanying them to the gym if they fancied doing a bit extra on the other nights of the week. Rather more illustriously, he was also involved with the football club. I'm sure he was brilliant with the big names on the other side of town – to Blackie, working at St James's Park under Kevin Keegan was the next best thing to wearing the black-and-white strip of his dreams – but he had a real soft spot for rugby. In the early days of our association, I don't think he could quite believe he was handling a bunch of athletes who played a supremely combative sport with such skill, and who actually enjoyed putting themselves through hell in the gym. The players were still up for a good time when the moment was right – we weren't long out of amateurism, after all – yet we were also tough and professionally minded, with a win-at-all-costs mentality he found inspiring. Blackie being Blackie, he gave people the impression that all his Christmases had arrived at once, even when he was not feeling on top of the world. But in our case, I think he genuinely felt it. He saw an opportunity to drive a culture of success, to sustain it and improve it over time.

It struck me very quickly that Blackie would be good for Jonny – not that you had to be Sigmund Freud to work it out. This is not to suggest that he wasn't good for the whole squad: pretty much without exception, he was of significant value to everyone. He cared deeply about the team, but he cared more for the individuals within it. Inga Tuigamala was a different character to Dean Ryan, who was different to Pat Lam, who was different to John Bentley. Blackie understood that while these people were on the same side, the common denominators ended there: the psychological support he offered was targeted like a laser beam. But it was his work with Jonny that defined him in the collective mind of the rugby public – and, as things developed, of the far broader spectrum of the English sporting public as a whole.

What did he find when he first started working with Jonny? A fixation with being the best that he had not previously encountered in any area of sport. Like Blackie, I felt I understood what lay at the root of it: I'd worked hard at my own rugby because I too was a perfectionist. But there are different degrees of perfectionism. Not all perfectionists chase their optimum level of performance in the same way, still less take their pursuit to the same limits. I felt I'd gone about things with a sense of balance, born of the understanding that all sports – rugby, cricket, football, whatever – are inherently imperfect. In order to stop yourself going barmy, you have to acknowledge that while you're always aiming to play a fault-free game, and that just occasionally you might go somewhere close with your kicking or your passing or your tackling, there will inevitably be something you might have done better, if only marginally. Jonny found that sense of balance elusive, I think. He was always at risk of becoming consumed by the pursuit, of driving himself deep into a place where he was attempting to achieve the unachievable. Blackie might have been taken aback by his intensity at first, but he made it his job to connect with him, encourage him and, in a way, protect him from himself. Was he a crutch for Jonny? I believe there was an element of that, yes.

Poor Blackie. The man had energy, oodles of it, but as Jonny cemented his place in the Newcastle line-up, closed in on the England squad and ultimately became Clive Woodward's long-term No. 10, there were barely enough hours in the day.

Jonny was incapable of accepting that something was beyond him. When he arrived at the club with all his skills, two things were evident: that he could not be counted among the fastest midfielders of all time – I knew what that felt like myself – and that he did not possess the kind of step that might minimise the effects of this lack of extreme pace. Yet the moment Jason Robinson, who was seriously rapid and whose step was a gift from the gods, came on the scene, Jonny said, 'I want to do that', and worked himself to a standstill in an effort to emulate the master. It happened time and again, and whenever Jonny set himself a fresh target, Blackie would have to devise programmes and training sessions to that end. And as he was the kind of man who would happily respond to each and every request for help, he would drop everything if Jonny phoned him and said: 'I need to do some stuff. Can you meet me in the gym?' He would then spend as much time as needed catching Jonny's passes or fielding his kicks, or watching him play keepy-uppy with an oversized tennis ball for hours on end. Hours! This is no exaggeration. Jonny's control was such that when he was at the peak of his powers, it was impossible to tell if he favoured his left foot or his right.

If it was easy to detect an air of compulsion about him right from the start, it was also blindingly clear that he had a proper future in front of him. Jonny trained the house down, each and every time he set foot on the field; he performed well at second-team level, stepping up to meet the physical challenge of tangling with the grown-ups while bringing his burgeoning skill set to bear on unsuspecting opponents; and he listened intently whenever the senior players were discussing tactics and strategy in his presence. He was not the greatest talker in those early days, but he was almost sponge-like in his capacity to absorb information. It was the correct approach. There were so many big characters in our Premiership squad, so many international-quality players who had been there and done quite a lot, that no 18-year-old kid concerned for his own wellbeing would have dared open his mouth. Was he in awe of some of these people? For sure. And who could blame him?

These were wild times, the pioneering age of panning for rugby gold, and he'd pitched up in a squad that had been thrown together by two or three hard-bitten old amateurs working off the back of a fag packet. Dean, Steve, myself – we'd all accumulated some leadership experience during our spell together at Wasps. If our dear old head coach, Rob Smith, was ever late for training – hugely committed to the club, he would make the long trip up to Sudbury from his home in Bristol in a van that was not immune to the occasional mechanical breakdown, and he also had a dog that sometimes went missing – we'd fill in until he arrived. The three of us had therefore been responsible for quite a bit of the organising down in London, and we continued in that vein at Kingston Park. The added ingredient was Blackie, who was sometimes a firecracker in the middle of a bonfire – he was no stranger to anger – but frequently managed to turn himself into a bucket of cold water.

At some point in any given squad session, Dean might say 'I've spotted this and it's not happening again, so I'm going to keep people on the training ground until it's sorted.'

And Blackie would be the first to say 'That's not a good idea' – quite a brave stand when Dean was on a mission.

Dean: 'I'm doing it anyway.'

Blackie: 'We need our best people fit and on the field, not sitting in the stand watching a game they're meant to be playing.'

At which point Steve and I would say: 'Actually Dean, he has a point,' and hope for the best.

It was a matter of conflict resolution, and Blackie knew how to do it.

And then there were those charabanc trips: six hours to wherever, play the game, six hours back with a crate of beer via the Wetherby Whaler chip shop just off the A1M, across the Tyne Bridge, into the middle of town to drop off the drinkers who fancied a proper night out (even though the night was already two-thirds over) and then round the houses to deposit those far too knackered to contemplate anything other than bed, even though sleep would be interrupted by the horrifying thought of it all beginning again on the Monday morning. It was a crazy world: part professional reality, part amateur spirit; part serious, part not-so-serious. For a teenager, even a teenager as driven and focused as Jonny, to find himself in such company must have been quite an eye-opener for him.

We always had the idea that if his rugby developed quickly enough and his form merited it, we would work Jonny into the side as a centre and let him spend a couple of years standing next to the old man in the No. 10 shirt, learning the ropes. It was not a revolutionary plan – most famously, the ground-breaking 1984 Wallabies played a young Michael Lynagh at centre, outside the magical Mark Ella – but it was a good selling point in our discussions with the Wilkinson family. There might have been moments in the first few weeks of his debut season when Jonny, holed up in his less-than-palatial temporary home in the West End of town, wondered if and how things would come together for him, but by the late autumn of 1997 he was a regular feature on the first-team bench and was pressing hard for something more in the early months of the new year. After our defeat at Richmond, I decided there was a strong case for change: the Tait–Tuigamala partnership was showing slight signs of drift and I felt Jonny would bring something fresh to the midfield – not least by giving us a left-boot option at 12, which is always a bonus when you have a right-footed kicker at 10. It was a bold call, saying to a series-winning Lion and a celebrated All Black-turned-Samoan folk hero that one or other of them would be missing out on a start, but it was also the right call. Jonny started eight of the last nine Premiership fixtures and was an ever-present figure across the six-match run-in. In fact, he played all but 20 minutes of those half-dozen games. You don't get to do that with a title on the line unless you deserve to be there.

Not that things went to plan immediately. We won tight contests at home against Saracens and Wasps immediately after introducing him into the starting line-up, but by a strange quirk of the fixture list, we had to travel back to the same opponents almost immediately and lost on both occasions. It was Lynagh, now at Saracens, who did for us in the first game, nailing a very late drop goal to pinch it 12-10 in front of a 20,000-strong crowd at Vicarage Road in Watford. Three days later – yes, three – we found ourselves in an equally fierce scrap with my old club at Loftus Road in West London, a contest that went against us 18-17 when I missed with the last kick of the night. I would have liked to have blamed our defeat on the kid next to me, but it was difficult to see how I could get away with it under the circumstances . . .

Life at the top was seriously claustrophobic by now, so it came as quite a relief when we put seven tries past Bristol in our next game before taking two matches, big ones against Leicester and Bath, to Gateshead International Stadium. This was another risky move, but we felt we could double our crowd by shifting to a bigger venue, and so it turned out. By pulling away from the Tigers in the second half and then chiselling out a five-point victory over the West Countrymen in a match that was some way short of the most enjoyable of my career (foul night, foul tempers, grim all round), we set ourselves up for a tilt at glory at Harlequins, the final match of the campaign.

Aboard the charabanc once again, we headed south 24 hours early: Newcastle United were playing Arsenal in the FA Cup final – one of the last at the old Wembley – on the Saturday; our game across town at The Stoop was scheduled for the Sunday. The football did not go at all well, even though us rugger types were cheering on our brethren from the stands, goals from Marc Overmars and Nicolas Anelka condemning the Geordies to a 2-0 defeat. We didn't see Sir John afterwards. Maybe he was in the dressing room.

We did see him at Quins the following afternoon, however, and he was in high good humour, with every reason. We turned in a champion performance in front of something close to a full house, rattling up half a dozen tries in a 44-20 victory. I converted four of them and banged over a couple of penalties for good measure, a decent enough way to crown a season-long effort that had been exhausting and exhilarating in equal proportions. The aftermath is still a blur. I don't remember Sir John occupying a seat on the bus as we headed for home, and I don't remember spotting Jonny either. The one thing I'm sure of is that we went drinking with the Quins boys in Richmond before boarding the coach at two a.m., joined by a large band of supporters who had been celebrating with us in the pub and had no alternative means of finding a way back to Newcastle in time for work.

By this time, Jonny could legitimately describe himself as an England international. Clive Woodward was so short of wings for the final Five Nations game with Ireland at Twickenham, he had run Mike Catt in the position – and then thrown Jonny on as his substitute with two minutes left on the clock. Strange to relate, we'd pulled a similar stunt about three weeks previously, during our defeat at Richmond. I'm not sure Jonny was particularly comfortable in his unfamiliar role at the Athletic Ground and I can't believe he was any happier when asked to stage a repeat performance in front of a rather bigger audience a couple of miles along the road. Being the very epitome of the process-driven player, the thought of standing fully exposed in the wide-open acres without much in the way of straight-line speed to cover his modesty must have been unnerving. At least it was over almost before it started. Unfortunately, his next experience of international rugby, in his proper position, was a whole lot tougher for a whole lot longer.

That summer, England travelled to the southern hemisphere to follow an absolute brute of an itinerary: a Test against Australia in the great Wallaby stronghold of Brisbane; five games in New Zealand, including two nice little runarounds with the All Blacks and three horrible midweek fixtures, one of them a real mugger's alley meeting with the murderously competitive Maori in their Rotorua heartland; and, just to rub it in, a set-to with the Springboks in Cape Town on the way back to Heathrow. Clive selected a bare-bones kind of squad, missing more important players than it included, and Jonny was slated to start at No. 10 on opening night in Queensland. He missed the sticks with a couple of straightforward penalty shots early on, but given that the tourists went on to lose 76-0, those errors did not amount to much. Of more concern was the effect such a desperate defeat might have on an ambitious, hugely talented yet soft-boiled newcomer only just old enough to vote.

To his great credit, he showed no signs of terminal damage when he returned to Newcastle for our defence of the title, and as he grew into his rugby at both club and Test level, it started to dawn on everyone at Kingston Park that we had among us a player for the ages. He was still a couple of years shy of the first of his high points as an outside-half – it was not until the early 2000s, the period of England's long unbeaten run and the eventual World Cup triumph, that he raised himself to his full height as a No. 10 and established himself as the best in the sport at that time – but I was in no doubt that once he felt secure in the structure around him, he would wield an ever-greater influence with both club and country.

In England terms, I think his personal journey through the 1999 World Cup and the years immediately following reflected the journey of the team, which in turn was probably a reflection of selection, which was more than a little fluid, to put it mildly. Jonny was not naturally suited to chaos: he was not a Catt or a Robinson, who could make high-risk plays off the cuff in situations of dynamic change and get away with it; he was not a Will Greenwood, whose instinctive grasp of rugby in all its facets and understanding of the range of possibilities available in any given circumstance made him such a valuable member of the World Cup-winning side. With Jonny, the instinctive side was nowhere near as marked. The things he brought to a team – and he contributed as much as anyone of his generation – were the product of an ingrained commitment to improvement. He was a professional, skin, pips and core: more than any of his peers, he spent countless hours reflecting on his game, recognising the flaws within it, identifying new skills that might counterbalance those weaknesses and then perfecting them until he could execute them as well as anyone, if not better.

Talking of reflections, I sometimes think that the England team of 2003 came to reflect Jonny. Fundamentally, the World Cup triumph was hewn out of a collective efficiency born of hard work and the gradual piecing together of a formula so cohesive that it ultimately overpowered all the other leading teams in the tournament (none of whom, if we're honest, were anywhere near a peak when it really mattered). It came to them through trial and error: lots of trials, lots of errors. Clive's original notion of how he wanted England to play – and still encouraged them to play, especially in the early 2000s when Brian Ashton was on board as attack coach – seemed to have little in common with the rugby his players produced at the World Cup. In the end, he moved his philosophy towards the players rather than move the players towards his philosophy, and as a result, England became incredibly difficult to beat.

The England who chucked away a Grand Slam by spurning kicks at goal against Wales at Wembley; the England who messed up another clean sweep by trying to ping the ball around like the Harlem Globetrotters in the middle of a Murrayfield monsoon – that England was dead and buried come the autumn of 2003. By hook or by crook, they had taken themselves to a place where everyone knew what was happening pretty much all of the time: the place where teams need to arrive if they're to stand a chance of fulfilling their potential. For England, read Jonny. To him, uncertainty was anathema. Playing it off the top of the head? No thanks. Going off-script simply didn't work for him: why have a game plan if you're going to unravel it? What he needed was a direction of travel to follow and a set of parameters to perform within. Give him that and he could make all manner of wonderful things happen over the course of 80 minutes. If the magic he had was different to Robinson's or Greenwood's, it was still magic.

Of course, it is a matter of fact, if not of formal record, that the World Cup campaign was rather less comfortable for England than the performances over the previous year suggested it ought to have been. They'd defended Twickenham's honour through the 2002 autumn internationals, beating the All Blacks and the Wallabies by the combined total of four points – Jonny completed a full house of try, two conversions, a drop goal and three penalties against the New Zealanders, and followed that up by putting 22 points past the Australians with the boot – and then smashed an overtly physical but otherwise hopeless bunch of Springboks by the record margin of 53-3. They had then completed a long-awaited Grand Slam, clinching it with a second-half demolition of Ireland in Dublin, before doubling down on their victories over the Antipodeans by squeezing past the All Blacks in Wellington and running rings round the Wallabies on a blistering night in Melbourne. Here was a side at its peak, with a World Cup on the horizon. What was there not to like about England's chances?

Without wishing to sound wise after the event, I feared for them just a little before they departed for the tournament because I too had played rugby under the dead weight of expectation – of assumption, even – and I knew how debilitating it could be. The burden can feel terribly heavy, to the point that you find yourself wishing you were anywhere else but on the field of play. I can remember being involved in really big games, leading after 70 minutes and thinking: 'I'm not enjoying this one bit. It's almost painful. Why doesn't the referee just blow his bloody whistle and let me out of here?' I could sense that this kind of negativity might afflict England during the competition. For one thing, they had already climbed the mountain on that summer tour. To go back out there and climb it all over again was asking plenty. More than that, they would be travelling as the number one side in the world and strong favourites for the title. That brought its own problems, not just for Jonny but for the vast majority of the party. Big problems. Who among us can say, hand on heart, that Clive's team played really well over the course of those six weeks or so? Did they perform against the Samoans? Against the Welsh? Hardly.

The BBC had asked me to be part of the radio commentary team for the semi-finals and finals and had booked me on the long-haul flight to Sydney. At half-time in the quarter-final with Wales, I texted Ed Marriage, the rugby producer, and asked him if I should cancel the ticket and unpack my bag. We were being completely outplayed and the game was on a knife-edge, at best. To be frank, there was only one team in it and they weren't wearing white shirts. Somehow England found their way out of the corner in which they'd trapped themselves, so out I flew. Jonny wasn't in a great place mentally when I arrived; as we sat down for a coffee, I thought he seemed . . . well, lonely. He'd been in Oz a long time and had very much kept himself to himself the whole trip. He didn't want to go out socially, he was almost certainly practising more than he needed to, and without Blackie around to keep him energised, he was struggling to cope. Ironically, it was the semi-final – the game traditionally considered to be the most brutally nerve-wracking and tension-filled of all – that helped him rediscover some equilibrium at the crucial moment. For Jonny, process was a comfort blanket: it was something he understood and felt he could turn to in times of need. If you're an introvert by nature, a self-isolator to the point of being hermit-like, there are precious few avenues of escape when things stop going to plan.

As Jonny was also the fulcrum of the England team, the goal-kicker and principal provider of points, the 'match-winner', the glamour kid (however much he loathed that image), there was no avoiding the adulation of the England supporters or the bile from the Australian media. If ever he needed process, it was then. And because the Sydney weather had turned foul, the match itself became an exercise in process.

Under such conditions, the game was bound to be tight and narrow and territorial, with minimal opportunities for outbreaks of attacking extravagance from the French backs or flashes of individual genius from a Frédéric Michalak or a Christophe Dominici. Right up Jonny's street, in other words. I hesitate to say it was a wet walk in the park for England, as no game of such magnitude can ever be entirely simple but, as semi-finals go, it was as blissfully straightforward as it could possibly have been. The French were all but beaten when they walked on to the field (one of their players had been heard to say the previous evening 'If it rains tomorrow, we are dead before kick-off', which told us all we needed to know), and while they scored the only try of the night from a cock-up at the line-out, they finished a million miles short of Jonny and his 24 points. He played exceptionally well that night and I was proud of him, partly because I knew how much he'd given of himself in pursuit of the prize over many years and partly because I knew where he'd been psychologically just a few days beforehand.

When we met for coffee again in the week of the final, some important pieces had fallen into place for him. I'd seen him in happier states, but he seemed better than he had been a week previously. And as the great showpiece collision with the Wallabies took its course, the core elements of his game – kicking and defence – rose to the surface. Looking back, the biggest kick from the team's point of view wasn't the famous drop kick off the wrong foot that actually won the match, but the 50-metre-plus, wide-angled penalty he struck in the first half of extra time. There was God's amount of pressure on him at that stage, but he found the inner strength to land as monumental a shot as I can remember seeing. But for that goal, Jonny's future might have been very different, and English rugby might have found itself on another road.

So what could we expect from Jonny when he returned to Tyneside? The place was agog with excitement, which made a nice change: since the title-winning year in 1998, things had grown a little more difficult for everyone at Kingston Park. A fairly elderly group of players had largely gone their separate ways; we'd missed out on a big-time European campaign because the English clubs had boycotted the Heineken Cup in 1998–99 (rugby is never far away from a political flare-up), and our Premiership performances had fallen victim to the law of diminishing returns. Eighth, ninth, sixth, sixth, tenth . . . It hadn't quite been what we'd had in mind when we'd arrived back in Newcastle with the trophy after that joyous victory over Quins at The Stoop. With Jonny back, the local rugby public believed things would take an upward swing.

A week after the drop goal in Sydney, we had a home game with Wasps. Jonny was there, as was the almost equally star-dusted Lawrence Dallaglio, parading the Webb Ellis Trophy in front of a capacity crowd. You couldn't have wedged a cat into the ground that day, let alone another human being. The place was rocking, and the supporters assumed there was more to come.

Except Jonny was injured. An awful lot has been said and written about the neck problem that gave him so much unmitigated hassle, and cost him so much of his prime, in the years following the World Cup final, and it is indeed true that the heavy hits he took in that climactic game set him back. But the truth of the matter was that he had been suffering in that area before the tournament, and I'd have been more than a little concerned for his welfare even if he hadn't taken such a battering from the Wallabies.

We gave him a decent break, then picked him for our home game with Northampton three days after Christmas. There was another 10,000 crowd, double our usual gate, and they sure as hell weren't there to watch the props scrummage. Jonny had been on the field for 50 minutes when he tackled Jon Clarke, one of the more substantial backs on the circuit, and stayed down. It was the trigger for an injury run that would, give or take the odd brief period of relief, lay him low all the way through to the next global tournament in 2007.

It was desperate news for the club: a fit Wilkinson at the height of his fame would have made a massive difference to us, both as a team and as a business. But if it was sad for Newcastle, it bordered on the tragic for Jonny. The injury, and all those that followed, left him in every kind of emotional strife. I really felt for him. Each week, there was the same question from the media about Jonny's potential return date;

each week, I gave the same shoulder-shrug of an answer. It wasn't that I was being deliberately evasive; it was simply a case of not knowing. Every so often there would be a promising sign, but even when Jonny did find his way on to the field, after weeks and months of gut-busting work in the gym, something else would go wrong. There were knee problems, there was a haematoma in his right arm, there were groin issues, there were shoulder complications. In 2004, we flew to Japan for a pre-season tour, all expenses picked up by the hosts because Jonny was as much a superhero there as he was in England, if not more so. What happened? He went down with appendicitis. Instead of playing a game of rugby in Tokyo, he spent his time in a downtown hospital. All things considered, it was beyond miraculous that he recovered sufficiently from such enduring trauma to achieve what he did in the second half of his professional career.

How good was he? Where do I place him in the outside-half batting order, to mix up my sporting loves for a second? It's difficult – terribly hard, in fact – to pass judgements on a cross-generation basis, for the simple reason that the game Jonny ended up playing was somewhat different to the one Michael Lynagh played, and barely recognisable from the one mastered by Barry John and Phil Bennett in the Welsh glory days of the 1970s. I don't think he was as naturally gifted in terms of the sport's unmeasurables and abstractions – the peripheral vision, the instant identification of space and maximisation of it, the instinctive recognition of an attacking opportunity – as some of his fellow front-rankers, but then, the perfect 10 never did, and never will, exist. From the recent English perspective, you might go close if, say, George Ford, Owen Farrell and Jonny were rolled into one. There again, you might as well go hunting for unicorns in Twickenham High Street. For many observers, there were moments when Dan Carter, the World Cup-winning All Black, justified the use of the p-word as he put the British and Irish Lions to the sword in 2005 but, let's face it, those Lions were a long way short of top quality, in performance if not in personnel.

Jonny had all the skills, technically speaking, but he didn't quite have all the gifts – not in the way Carter had them. His best international rugby was produced when he had someone outside him who could assist with the decision-making and release the pressure valve, physically and especially mentally. I'm talking here about a Greenwood or a Catt. Having a second game-manager in close proximity allowed Jonny to be absolutely brilliant at what he knew he was good at: his kicking – at goal, cross-field, chipping, long out of the back field – and his passing of all ranges, together with his tackling, which was as watertight as any we've ever seen from a player in his position. In terms of what we call 'closed' technique, he was as good as anyone because he practised and practised and practised. The 'open' techniques of instantaneous decision-making? He wasn't always the best there: on a bad day, he could disappear up cul-de-sacs and lose himself in blind alleys. In the art versus science debate, which I'm not sure is wholly relevant to the discussion about Jonny or any other player, it's clear that he would be placed – and would probably place himself – in the techno-mechanical column.

Not that this detracts for one second from Jonny's human side. His career statistics – tournaments won, points scored, tackles completed, contributions made – may have rugby's mathematicians salivating, but I prefer to see his career through the prism of commitment, determination and self-sacrifice. Those qualities are the product of a beating heart, not a machine. I don't think I've ever seen anybody push themselves to the limit in pursuit of a set of objectives, both personal and team-wise, in the way he did. To get back to international standard after such an injury blight, play at the 2007 World Cup and almost pull it off – that was truly remarkable. Had the tight final with the Springboks taken a slightly different turn, it would have been the most extraordinary rugby story of them all. And then to win selection for the 2011 tournament, and then to drive Toulon to three European titles, and then to bow out in Paris having won the French championship, with the local crowd singing 'God Save The Queen' – with their history! Fiction wouldn't dare to go there.

If he was not, probably by some distance, the most naturally blessed player who ever appeared at No. 10, has anyone ever squeezed more from their reserve of talent? He showed a fidelity to his work that went beyond the call of duty. Way beyond, to the point where it became obsessive. Correction, it went well past obsessive. He'd be the first to admit it. For him and his closest allies (his parents and brother, who were astonishingly supportive of him every step of the way, and family outsiders who became insiders, like Blackie), managing that obsession and ensuring that it didn't become wholly destructive was hard work. In the end, though, he emerged with an awful lot of the rewards he deserved.

I don't know the answer to this, and I don't suppose Jonny does either, but what would have been his response if, when he was ten years old, someone had told him: 'This is what it will look like when it finishes – these winner's medals, the World Cup final drop goal, the late re-flourishing in France. And the price you'll have to pay for it is going to be this big. What do you say?' My hunch, for what it's worth, is that he would have replied: 'Yes, the price tag is big. Terribly big. But I'll still pay it.' Are we talking about a Faustian pact here? Quite possibly, but he wouldn't have wanted it any other way. For most people, there is no pact to be made: there are untold millions who love the thought of the outcomes, but how many are prepared to meet the asking price, or are even capable of meeting it? We are all creatures of our limitations.

5
FIGHTING FOR THE FUTURE

I BROKE LONG BEFORE Jonny did, orthopaedically speaking, but I had an excuse: I'd been playing top-level rugby for a good 16 years, not six, and had the scars to prove it. You reach a point where the next scar becomes one too many. I knew Jonny would be heading off with England to the 1999 World Cup in France – if he'd been some way short of ready for the 'tour of hell' on the far side of the planet a year previously, he was a justifiable pick for the rather more important trip across the Channel – so I made the decision to start the 1999–2000 Premiership season myself, even if I suspected that I might not reach the end of it. Leaving aside the fact that I was now nearer 40 than 30 and that my week was quite full enough without devoting precious hours to high-level training, my shoulder was comprehensively knackered.

Six dislocations in the space of two years provided me with a massive clue that my time was just about up, but I felt I could at least provide cover for Jonny until he returned from international duty. I was a little nervous, however, because with each dislocation, the pain had intensified. Dislocations are not great news irrespective of where on your body they might occur: anyone who has looked down to see a finger pointing at right angles to the rest of his hand will be aware of the discomfort involved. Busted shoulders are at the higher end of the agony scale and, while I hadn't undergone surgery at any stage as the medics always seemed to find a way of putting things back together without reaching for the general anaesthetic, the last one did for me. My final game was against Gloucester at Kingsholm, where opponents never anticipate a sympathetic reception and rarely feel the need to revisit their expectations. We lost 31-16 and some of the more loquacious Kingsholmites could be heard claiming that the ferocity of the Gloucester defence had put an end to my career. A good story, but a wrong one. The dislocation occurred a couple of days later, when I hit a tackle bag in training. I hesitate to suggest that colliding with a slab of foam rubber was more perilous than running into a Cherry and White forward but the facts speak for themselves. So there I was in the ambulance, on the way to hospital for some much-needed oxygen, when I asked myself the pertinent question: 'I'm 36. Why am I still doing this?'

It had already been one heck of a calendar year, beginning with a bolt from the blue in the form of Sir John Hall's declaration of retreat from everything to do with the union code. For four years, he had been in the thick of professional club rugby, warts and all. He'd helped spark the thing into life with his initial purchase of a struggling set-up on the far frontier of the domestic game; he'd relished his role as agitator-in-chief on the politics front, working his way under the collective skin of the Rugby Football Union and driving that august organisation to distraction; he'd patently enjoyed the travelling, the camaraderie and the banter on the team bus; he might even have seen the funny side of losing some of his small change to Doddie Weir and Gary Armstrong. But suddenly, he wanted out. He'd lost millions on the Newcastle venture, but I don't think the money was the overriding issue. The main problem, I believe, was that he couldn't see where club rugby was going – that he simply couldn't see a way through. He'd won the Premiership title, he'd had his fun, and now he had decided that the sport was in no position to deliver what he thought it might when he bought into it. 'You have two months,' he told me, just after Christmas 1998. 'From the end of February, I won't be putting in another penny. I love the club and I don't want to see it collapse, but you need to find yourself a new owner.' And that was it. No more money. It may have been a wholly pragmatic business decision from his point of view, but it was a scary one as far as I was concerned.

If I'm being honest, it felt like a blow upon a bruise. We had already slipped off the summit to which we'd ascended the previous season, largely because we could not compete in the Heineken Cup, thanks to a boycott of the competition staged by all the top-tier English clubs, together with Cardiff and Swansea, the most powerful teams in Wales. I knew the squad would soon need renewing, that I'd have to replace the old workhorses I'd brought in to do the heavy lifting at the start of Sir John's stewardship, by bringing through some home-grown talent from the north-east and beyond, but I'd been equally convinced that the Premiership-winning side had another strong year left in it because of Europe. We would have had a proper shot at the likes of Toulouse and Brive and Stade Français if we'd been given the chance, but the latest outbreak of strife between the club movement and the governing classes had denied us what we'd craved. Instead, we found ourselves underperforming in the domestic league and filling our spare time with so-called rebel games against the two bands of Welsh refuseniks: an exercise in gesture politics that was always going to end shambolically.

I felt even worse about things when Ulster and Colomiers contested the European final shortly after Sir John announced that he was cashing in his chips. Colomiers? During my time playing for Toulouse, they were a bunch of upstarts from the local suburbs. Their presence in Dublin for the showpiece occasion reinforced me in my belief that we could, and quite possibly would, have been there ourselves.

Out of Europe and soon to be out of funds . . . We were in serious difficulties and in dire need of a speedy solution. I worked closely with Ken Nottage, then the chief executive of the Newcastle Sporting Club and its four components – the football team, the ice hockey and basketball, and us. There was no sugar-coating it: we were being cut loose and it was his job to get rid, albeit as painlessly as possible. The two other minority sports would go the same way soon enough. How quickly the dream of an English Barcelona had faded to nothing.

With Ken tearing around trying to attract an individual buyer or piece together a consortium with sufficient spare pennies between them to drag us out of the mire, I was doing everything in my limited power to help. It was one of the more testing times of my life: I'd invested a lot of myself in building up the Newcastle operation but, more than that, I'd persuaded a lot of good people to join me in the venture.

It was at this point that I found myself talking to a similarly afflicted director of rugby in the shape of John Kingston, who had been my first captain at Cambridge University. John was now my opposite number at Richmond, our dear friends and deadly rivals from second-division days. A more alarming similarity was that he too had seen a backer close his wallet and head for the door – in his case, a multimillionaire by the name of Ashley Levett, who had made a fortune in copper trading and lived, when he wasn't enjoying the high life in Monaco, in a mansion just outside Winchester. (John would one day recall popping down to Hampshire for a meeting, being shown into the library and realising as he entered that this one room was significantly bigger than the entire ground floor of his own house.) We spoke regularly on the phone as our respective crises took their courses, and the conversations had a bleak humour about them. 'Our owner's pulled out.' 'That's funny: so has ours.'

Fast-forward a few days. 'If we don't find someone to take the thing on over the next couple of weeks, we're stuffed.' 'Same here. Let me know how you're getting on.'

Spool forward again. 'How's it going?' 'Not so much as a nibble. You?' 'It's not looking good.'

As luck would have it, Dave Thompson materialised at precisely the right moment for Newcastle. (Richmond, sadly, would not be blessed with such fortune.) Dave was a Geordie born and bred who had made his pile in information technology, sold up and returned to Tyneside to enjoy a comfortable retirement. He was a passionate rugby man, enthusiastic enough about the game to have spent time coaching the kids' section at the old Gosforth club, which had floated back into the amateur world after the establishment of the professional set-up, playing their rugby on the university pitches adjacent to Kingston Park. When I spoke to him for the first time on the phone, he was clear about his interest in succeeding Sir John, and when we met face to face at the Wheatsheaf pub on the road to the airport, he said: 'I love my rugby, I have a few quid and I want to help.'

I swallowed hard. Word had it that he was a 'tens of millions' businessman, rather than a 'hundreds of millions' tycoon.

'Well, you'll need a few quid,' I replied. 'It's costing the best part of £2 million a year at the moment.'

I remember wondering if he was for real and I'm pretty

sure I asked him, as gently as I knew how, if he was crazy. He said he enjoyed a challenge. And so it happened, fast enough for us to avoid the kind of cash-flow chaos that would reduce other clubs to rubble. Sir John was only too pleased. 'Great, you can have the club,' he told Dave. 'We need a monetary value to make this right and proper, so give me a pound and I'll take care of the debts. I don't want to see it die on its feet.'

The word 'relief' does not even begin to describe my reaction to these events, but other thoughts were swirling round my head at the same time. We hadn't simply dodged a bullet, we'd somehow avoided an entire barrage of heavy artillery. Good for us. But the experience left me with a very clear impression of just how fragile professional club rugby had become. It wasn't merely a matter of Newcastle and Richmond: everyone was losing an absolute shedload of money.

We were all spending too much on players, our squads were too big, there wasn't enough cash coming in. In short, playing budgets were completely out of kilter with the scale of the business as it existed at the time. Yes, the crowds were growing larger, but gate receipts were not generating nearly enough to provide stability. At the end of Year Four of the professional revolution, we had saddled ourselves with a model that could not have been more flawed had it been constructed by an innumerate child with a broken abacus. It was a horrible situation. On the club side, management people were racing about frantically trying to find ways of halting the downward spiral; on the governing body side, the traditionalist types who had never wanted the sport to go open in the first place and who certainly hadn't welcomed the likes of Sir John Hall into the game were in 'I told you so' mood. Relations that had not been the best to start with deteriorated as the financial pressures intensified. It was pretty obvious to everyone involved that English rugby would soon be at war with itself again.

Over the previous couple of years, while Sir John the agitator was in full swing, I'd played only a bit-part role in rugby politics. I'd been close to the odd skirmish, but quite a distance from the really serious battles. This was about to change. Within a few months, I would find myself at the epicentre of the debate over the future of the game in England, performing a role that would take up an awful lot of time and energy – so much of both that, had my shoulder not given up on me, I'd have had to retire from playing anyway. Having thoroughly enjoyed the Wild West stage of my spell with Newcastle, I was now heading into a kind of rugby dystopia – a place where creative ideas would be put forward and rejected, where strange alliances would be formed, where motives would be questioned and good works undermined. To many people, the politics of the game is about as tantalising as a bowl of cold cabbage. To those directly involved, it can be testing, frustrating and all-consuming in equal measure.

One of the most prominent figures in this new drawing of the battle lines was Tom Walkinshaw, the racing driver and team owner who had bought himself a majority shareholding in the Gloucester club in April 1997. He was behind the idea of a British League, which he was confident could generate some proper money; indeed, he was guaranteeing each club a £1 million participation fee – a level of funding that would have solved a lot of immediate problems – in return for the rights to the competition. It was bold and it was brash: had it happened, Tom could have been the Bernie Ecclestone of rugby union, which might have been his ambition from the outset.

He had support from a number of first-division teams, many of whom were failing financially and who found the prospect of an immediate seven-figure windfall deeply attractive. But the plan, which soon mutated into an Anglo-Welsh tournament offering £1.5 million per club, sent the RFU into another of its frenzies over who was controlling the game and for what purpose, and to be fair to the badged and blazered types at Twickenham, the Walkinshaw proposal did indeed amount to a breakaway, pure and simple. Suddenly, we were back in Kerry Packer territory, circa 1995.

The man staring at Walkinshaw across the no man's land of the rugby terrain was the chairman of the union's Club England committee, Fran Cotton. He had made no secret of his antipathy to the new breed of club owner, and had a good deal of public support from Clive Woodward, who, rightly or wrongly, had apparently chosen to take a political stand rather than concentrate on the day job he had only recently secured, which was running the national team. There was all manner of shenanigans going on, both between the rival factions and within the rival factions. There was a split on the club side, for starters: while a good number wanted to go with Walkinshaw, there were others, like Leicester, who found the idea of a breakaway too radical for comfort. Dave Thompson was in the Leicester camp. He had not been involved in Premiership rugby for long, but he recognised the threat of carnage when he saw it. He didn't really seem to get along with Tom Walkinshaw anyway.

Meanwhile, some of the heavy hitters on the union thought it might be a good time to play the 'divide and rule' card by putting central contracts on the table. They had missed the boat in 1995 and seen it sail out of the harbour but, as far as they were concerned, it had not yet disappeared over the horizon. If they could break the unity of the players by signing the best of them on RFU deals, they would be in a very strong position indeed. There were undoubtedly moments when those of us in the club movement thought: 'The only assets we have are the players and, if we lose them, this whole thing we're involved in could go up in smoke at any moment. And when the smoke clears, there'll be nothing left.' As it turned out, the players decided almost to a man that they would stand by the clubs, thereby ensuring that if we couldn't win the turf war, we wouldn't lose it either. It was from this stalemate that the first strides towards a Premiership salary cap arrangement were taken – a move as important as any in providing the clubs with a future.

To his credit, Fran decided it was time to rise above the bitterness and search for some common ground. 'We have to find a different way of doing this, of unpicking everything and putting something more sustainable in its place,' he said. I couldn't disagree. So when he set up a commission and asked me to take a leading role, I accepted. Why me? I'd been pretty open in stating that the system was broken, and I certainly felt that if English rugby didn't come together, we'd all be the poorer for it. Maybe Fran thought my range of experience was sufficiently broad to be of value. All I knew for sure was that this was a moment of extreme sensitivity. While the players had demonstrated considerable solidarity as a group and shown great loyalty to those owners and investors who had put their necks on the block in the financial sense, the uneasy truce could end at any minute.

So we set about trying to put together a bespoke English model that would offer viability and stability, both commercially and competitively. There were three sub-groups at work within the commission, which sounds like the worst kind of Soviet-style bureaucracy but which in reality worked reasonably well. Francis Baron, the first full-time chief executive of the RFU, chaired the committee with the most wide-ranging brief, and both Fran and I attended those meetings. There were also dedicated groups dealing with finance – far from unimportant, given the parlous state of the coffers – and academies, the development of which we regarded as crucial to the building of a professional game fit for purpose. There were some sharp-minded individuals around the various tables: Geoff Cooke, my old England coach, was involved, as was Mark Evans, who was then involved at Saracens and would go on to run Harlequins with considerable success before taking up a high-powered job in Australian rugby league.

We went through everything, twice and three times over: as an exercise in circle-squaring, it was as thorough and demanding as anything I'd previously encountered. As the final Club England Taskforce Report – the paper branded in the media as the 'Andrew Plan', on the grounds that I was apparently the overarching task-force chairman – stated in the first few lines of the introduction, we were offering a 'complete restructure of elite English club rugby' aimed at giving the clubs the security they craved while ensuring that the national teams at all levels, from seniors to Under-Whatevers, would prosper off the back of it. Talk about ambitious. Some of the issues we tackled, from promotion and relegation to franchising, to a complete redrawing of the northern hemisphere fixture programme, were as complex as could be, and I can't say I'm remotely surprised that the same subjects are still causing arguments today. And while I was getting my head around this stuff, I was still trying to manage a rugby club. Enough hours in the day? Nowhere near.

Even without the task-force work, I'd have had my work cut out. Transitional periods are always more demanding than those when things are ticking along nicely, and we were deep in transition. The title-winning team was breaking up – Dean Ryan was heading to Bristol, who had saved themselves from a Richmond-style collapse by attracting fresh investment from the new owner Malcolm Pearce, and were chucking money around despite the chill blasts from the prevailing economic winds; Alan Tait had shifted north to Edinburgh; we'd sold Pat Lam to Northampton, with whom he would soon win a Heineken Cup title. I can't say I didn't look back on what might have been when that happened. And if Inga Tuigamala was still on board, he was now costing us a fair bit more than we were receiving in return. (Although I have to say that when we somehow cobbled together a run in the Tetley's Bitter Cup and found our way into the 2001 final against Quins, the old man dragged us through almost single-handedly. Surrounded by a bunch of 19- and 20-year-olds, albeit ones as talented as David Walder, Jamie Noon, Tom May, Michael Stephenson and Jonny, he produced an unbelievable performance. He was past his physical peak by then and to make things worse, he was suffering from a trapped nerve in his arm and would undergo surgery the following day, but he brought his inner warrior to bear on events. It was one of rugby's great last stands and I treasure the memory.)

To make matters more interesting still, Dave Thompson was keen to make Kingston Park fit for the new millennium by drawing up plans for a major redevelopment. That meant counting the pennies we already had while sourcing a lot of extra pounds. There would be no quick, chequebook-driven fix on the team rebuild. Money was going to be tighter than ever.

Looking back, my close brush with financial reality clearly fuelled my aspirations on the task-force front. I was convinced of the need for some progressive solutions and I felt that, with a good deal of give and take, they could be more easily and lastingly achieved in partnership with the RFU than in confrontation with it. I took the view that there was more than one way to embrace radicalism – that the Premiership clubs could drive through a truly comprehensive set of reforms without breaking away from the governing body and throwing the entire sport into chaos – and that the senior figures at Twickenham, deeply attached to their own prejudices and privileges as many of them still were, would recognise the urgency of the situation and embrace something new. If it did not turn out that way, it was not for the want of trying.

The document we produced was wide-ranging, to say the least. Among the big-ticket items: a root-and-branch readjustment of the rugby programme in Europe, including the repositioning of the newly expanded Six Nations championship in an April–May window, thereby allowing the clubs seven months of almost uninterrupted rugby; a 12-team, 22-match English league tournament running from September to January, breaking only for a three-match series of autumn internationals; a nine-week stretch devoted solely to European competition; a guaranteed playing break of two months; a 30-match limit for Test players.

Among the even bigger-ticket items: the setting up of a joint union–club commission to manage the elite end of the game on a 50-50 basis; the establishment of this commission as a limited liability company subject to the appropriate body of law, rather than as an unincorporated entity subject to nothing more than the whims and fancies of the people at the top table; and, perhaps most significantly, a move towards a franchising of the Premiership sides, based on geographical spread and strict minimum criteria covering everything from administration and ground facilities to marketing and financial performance. Politics being the art of the possible, I fully understood that if some of this was achievable quickly, given the grace of God and a following wind, the rest of it depended on delicate negotiations with a number of other 'stakeholders', to use the modern word. Would the people running the Six Nations countenance a move to later in the year? Would the French, so deeply in love with a national championship stretching all the way back to the end of the 19th century, accept that the European Cup final, rather than their domestic version, should mark the climax of the club campaign?

As for the franchise idea . . . Let's just say I appreciated just how much we were asking of people. Generally speaking, the basis of such an arrangement is the scrapping of promotion and relegation for the duration of the franchise, always assuming the holders remain financially solvent. It's how

American sport works at the top end. Indeed, it's how most of the sporting world works: we Europeans, manacled as we are to the football model of ups and downs, are in a minority. The plan put forward in the document that would be tabled for approval by the RFU Council – yes, Will Carling's 'old farts' were still alive and kicking, if slightly more arthritically – was for a four-year franchise, but it incorporated some trendy new thinking designed to convince the ultra-traditionalist wing of the governing body that Fran and I were not Oliver Cromwell and Thomas Fairfax in disguise. There was an expansion clause – two additional franchises in the first five years, if the conditions were judged to be right – and the potential for a reintroduction of promotion–relegation play-off matches after two seasons.

However, the complications around the last move were considerable. Fran had been a big supporter of the four-team divisional championship that had died a death at the end of the amateur era, and liked the idea of an equal spread of Premiership franchises: three each in London and the South-East, the North, the Midlands and the West Country. I too felt that it was a way forward in terms of freeing the union game from its age-old heartland straitjacket and spreading it across the country. Unfortunately, we weren't drawing up these plans on a blank sheet of A4. The North was fine – Sale in Greater Manchester, Leeds in Yorkshire, my lot on Tyneside – and the West Country had a well-established 'big three' in Bath, Bristol and Gloucester, although we were uncomfortably aware that scores of thousands of wildly enthusiastic rugby followers in the far south-west would miss out under our system, at least for a time. Elsewhere, things were as awkward as could be. The problem in the Midlands was one of paucity: Leicester and Northampton were genuine rugby hotbeds, but that's where it stopped. The problem in and around the capital was precisely the opposite: Harlequins, London Irish, Saracens, Wasps: something – or rather, someone – would have to give.

The return of promotion and relegation was equally knotty: if we were to maintain the geographical element of the plan, a new franchise could only replace an existing one if both were from the same region. When I made the move to Newcastle, there had been no obstacles and no elephant traps: for the only time in my rugby life, I'd genuinely been able to make it up as I went along. There was no such luxury now that I was dealing with an entire sport rather than a single club. Who was it who said that 'hell is other people'?

Still, those of us who had worked long hours to produce the final document felt there was plenty to recommend it, and believed there was at least an even-money chance of it winning the full support of the custodians of the English game and being implemented – much of it without further ado, the remainder after cross-border negotiations with other governing bodies and club representatives.

The first stage of acceptance was completed when the RFU board voted in favour. The only Twickenham body in need of convincing now was the full union council. Off we went to the Rose Room under the south stand at Twickenham for the crucial meeting, which was where we hit the buffers. Some might say we hit the 'old buffers', but I couldn't possibly comment. After Fran presented the paper in his usual plain-spoken way, Jonathan Dance, the member for Berkshire, led a charge against it, arguing that promotion and relegation was sacrosanct – that without it, the union would be abandoning its long-standing commitment to the 'seamless' game, whereby every club in the country could advance through the ranks as far as their performances on the field permitted. I expected Fran to respond, to make a counter-charge, but he declined to do so. The document was voted down – not decisively, but by enough – and that was that. Why did Fran stay in his tent when the battle was there to be won? I cannot say. Maybe he simply didn't want to rock the boat. Maybe he felt he would isolate himself if he pushed too hard. All I can say for certain is that a huge amount of work had come to nothing.

Funnily enough, when I look at the paper now, my blood does not always run cold with the sense of failure. On the face of it, the club game was left in exactly the same parlous position it had occupied at the start of all this soul-searching: not only had the task-force plan bitten the dust, but Walkinshaw's deeply divisive push towards a breakaway British League had also fallen flat on its face. Cardiff and Swansea had scuttled off back to their masters at the Welsh Rugby Union, tails between legs, and the general financial outlook was far from favourable. Yet we'd come out of it with a salary cap agreement, the strongest possible indication of a collective realisation that we could not just carry on regardless in the face of overwhelming evidence that we were getting professional rugby wrong. You might say that the owners had finally woken up, to the overpowering aroma of coffee. They knew that there would have to be a serious attempt at stabilisation, with or without the active support of the union, if they weren't to leave themselves vulnerable to another assault from the RFU hard-liners, who could resurrect the central contract issue any time they chose. In that sense, I guess the so-called 'Andrew Report' was a turning point.

But there are other times when I think back on the episode and say to myself: 'You know, all we did was paper over the cracks.' Sure, we calmed the situation in political terms, laid down one or two fundamentals of the club game that are still in place today, and avoided the slide into all-out conflict that would certainly have happened if the Walkinshaw plan had gathered momentum and we'd staged our breakaway. Had the British League been launched, the RFU would probably have said: 'Fine. No one playing in that competition will be considered for international selection. Even if we have to lose every game for two years – even if we have to pick ten pensioners, four Under-18s and a packet of crisps for a home Test against the All Blacks, we'll bloody well do it. Why? Because you'll run out of patience before we do.' We'd also demonstrated that interested parties from across the elite game – RFU types, owners, directors of rugby, coaches, financial experts from both sides of the divide – could work together constructively. In the end, though, two facts were staring us all in the face. The clubs didn't have the money they needed, and the union didn't have central contracts. Which meant that at some point, sooner rather than later, it would all kick off again. Which it did, within five years.

I do not believe we will ever see a cricket-style central contract system in English rugby. Does it matter? Not really, provided the relationship between the governing body and the Premiership remains strong enough to deliver the behaviours the RFU wants from its clubs. The problem – and we are talking about a problem of mighty proportions – will come not when that relationship frays (it is never unfrayed, truth be told) but when it snaps. The bottom line is this: the interests of the RFU and the interests of the clubs are not aligned, no matter how much public relations spin is put on it when the two sides are just about getting along with each other. The only alignment is through the chequebook, and that alone is never the basis for a lasting marriage.

England is not New Zealand or Australia or South Africa; it's not even Ireland or Scotland. The way our rugby is set up, partly because of history and partly because of the mess-up in 1995, follows the European sporting model as framed by football. English and French rugby bear far more resemblance, in organisational terms, to German and Spanish football than to southern hemisphere union. Therein lies the challenge: to create a mutually beneficial system that allows both the club movement and the national team to thrive. Spain and Germany seem to have found a way forward in the round-ball game, not least in the production of locally developed players. They are miles ahead of English football in this regard: if there are one or two front-line Premier League clubs who can be relied upon to do their bit in encouraging home-qualified talent, it's not systemic. English rugby is closer to the ideal, but only because of two hard-won agreements between the RFU and the clubs over the last ten years or so.

But the fact remains that the only thing delivering behaviour is hard cash. The sole reason English clubs have lots of English players is that they're paid to have them – because every year, the RFU is giving them millions of pounds they would struggle to find elsewhere. They hate the idea of conceding ground, of allowing the union to buy them off in return for a kind of central system through the back door, and they cast envious eyes at the Chelseas and Arsenals and Manchester Uniteds of this world. The Football Association is in no position to buy behaviour for the simple reason that the Premier League teams have more money than the FA does. That is not the case in rugby. Yet.

One last word on promotion and relegation, the immovable object that prevailed over our less than irresistible force at that deflating meeting in the Rose Room. Where do I stand on the topic now? In the long years since I grappled with it as part of the task-force project, it has continued to plague the sport: had it been sorted at the council meeting in 2000, it would have spared me many hours of mental gymnastics in the decade and a half that followed. It is the devil's own job to find the optimum position because there are so many different sides to the argument, so many conflicts of interest. I've been round the houses on this for the best part of 20 years and, while there are some days when I'm as convinced as I was when we wrote the original report that a franchised system with a geographical spread is still the correct way forward, there are other days when I see only the objections. When the subject crops up, I'm always reminded of the old joke about the person who asks for directions, only to be told that he 'shouldn't be starting from here'. How many people involved in future planning in all walks of life would sell their souls for that elusive blank piece of paper, rather than one with a century's worth of history scrawled over it? It's not especially helpful to say 'this is what I'd do if only I could', but the temptation can be overwhelming.

If things had happened differently on the ground floor of the Twickenham south stand that day, what would have been the consequences? Clearly, I walked into the meeting thinking that acceptance would be the best outcome. But the best outcome for whom? The England team? The club players? The owners? The second-division teams? The county delegates? Would safe passage of the paper in 2000 have killed the Exeter story stone dead? Would it have consigned the Worcester story to the flames? Would these ambitious teams still have emerged as Premiership-quality sides and enriched the top division in the way they have over recent seasons? Exeter might have found themselves with a franchise at some point, but there is no guarantee they would have kept faith with their existing squad, as they did when they won promotion to the Premiership in 2010. They might have done what we did at Newcastle and bought themselves a whole new side.

Worcester might well have been awarded the third Midlands franchise, but had they been beaten to it by Coventry or Moseley or some new Birmingham-based confection, they might have disappeared off the face of the earth and taken a valuable rugby audience with them. Another thought occurs to me: if the Exeter and Worcester stories had never been written because of the adoption of the task-force plan, neither would those of West Hartlepool and London Welsh and one or two other clubs who overreached themselves and ended up in Nowheresville. I know this much: had the proposals won the support of the union backwoodsmen, English rugby today would be very different. What I don't know for sure is if it would have been better.

6

NO BRAVE NEW WORLD

IT USED TO be so simple: we trusted the BBC and the other major broadcasters, we believed what we heard on the flagship news bulletins and we took on board the stuff we read in most of the papers most of the time, even if there were one or two rugby journalists who were always capable of making the average player think twice about taking things at face value. There was no talk of fake news or alternative facts, and if anyone thought they were living in a post-truth world, they kept very quiet about it. Yet even back in the 2000s, some things were not quite what they seemed. The numbers may have been there in black and white, but they didn't always add up.

It is a matter of fact that the England team reached the final of the World Cup in 2007 and were therefore within 80 minutes of becoming the first country to stage a successful defence of the Webb Ellis Trophy. It was quite a feat, for few nations had ever put themselves within striking range of winning successive titles. The All Blacks won the inaugural tournament in 1987 at a canter, but they were knocked out at the semi-final stage four years later. Australia, champions in 1991, fared worse still next time around – someone dropped a goal to knock them out at the last-eight stage, but his name slips my mind – while the Springboks could finish only third in 1999 following their triumph on home soil in 1995. Had England managed back-to-back victories at the most exalted level of the game, it would have been one hell of a statement. Not least to the southern hemisphere, who would probably have considered themselves doomed and headed straight for the Kool-Aid. The last thing they would have wanted to see was England, the most heavily populated rugby-playing country on the planet, and by far the richest into the bargain, getting their act together on a long-term basis.

Yet the reality was that the SANZAR nations had little to fear: even though England made it all the way through to the last weekend of the 2007 jamboree in France while the New Zealanders and the Australians were kicking their heels at home (I still chuckle at the story of the planeload of rather presumptuous New Zealand supporters flying at 35,000 feet towards Paris for the semi-finals and final, only to glimpse their heroes on a plane heading in the opposite direction), it was clear by that stage that the opportunity created by the victory in 2003 had been frittered away. The people involved at the top end of the England operation were doing their best, but they were fire-fighting their way through one conflagration after another. This wasn't how it was meant to be when Jonny Wilkinson popped over his famous three-pointer in extra time and Martin Johnson reached towards the Sydney night sky with the golden pot in his hand, having received it from an Australian prime minister who would probably rather have thrown it at his head. Those were the days.

So what happened to the legacy? How did England contrive not to build something lasting on the foundations laid down by the Johnson vintage? I'm afraid to say that such failings have been England's default position right the way down through the decades. Leaving to one side the current bunch under Eddie Jones, who have shown some encouraging signs of turning into a genuinely top-of-the-range outfit, only once in my lifetime have I seen the national side scale a major peak and stay at altitude for a substantial period of time: in the early 1990s, when Geoff Cooke was coaching and Will Carling was skippering. If we look back on the Grand Slam-winning side under Bill Beaumont in 1980, we see a group of players, long on experience, who achieved critical mass in a single Five Nations championship. It was a 'one night only, never to be repeated' event. The following year they lost to both Wales and France; the year after that they slipped down a place to third in the championship table; the year after that they finished bottom of the pile. The best they managed in that tournament was a draw in Cardiff. Over that period the selectors picked three different outside-halves, three open-side flankers, half a dozen locks of contrasting styles and abilities, and could never quite settle on a front-row partnership.

You can see something similar in the fate of the stellar 2003 side, although they had been much more impressive in the build-up to their big moment than the 1980 boys. How quickly they fell to earth. After the World Cup victory, there were very few retirements: Johnson called it a day, which was far from ideal, and so did Jason Leonard, but as the venerable Harlequins prop had been in and around the Test front row since the summer tour of Argentina some 13 years previously, you could hardly blame him. They also lost the priceless services of Jonny Wilkinson due to that neck injury of his, which could now be categorised as chronic. Yet virtually all of the remaining 30 or so squad members were still up for selection and keen to capitalise on their unprecedented achievement. What did we see? A sharp decline of the most immediate kind.

England have almost always been unstable at international level: as a Test-playing nation, we have tended to build slowly through endless bouts of hitting and missing, enjoy a brief moment in the sun, and then disappear back into the darkness. And the most glaring reason for these failures to sustain our level of performance is clearly poor selection. It seems to me that coaches have either taken off on flights of fancy, or paid insufficient attention to the need for succession planning. As selection is the measure by which coaches must be judged – I wouldn't go quite so far as to say that everything else is just wallpaper, but the ability to pick the right side at the right moment is overwhelmingly the most important part of the job – this beggars belief.

When I was playing under Geoff Cooke, there was a clear selection policy in place. God knows, it was a relief. I'd been all too familiar with what felt like a 'pin the tail on the donkey' theory of team construction between 1985 and 1987, my early years in the England team: only six players from my debut against Romania were still in place when we bombed out of the quarter-final of the World Cup less than 30 months later. Some of the selection seemed capricious; some of it was worse than that. When Geoff was appointed head coach, he came in with a plan geared towards building a team capable of making an impact in the 1991 global tournament, much of which would be played on home soil in front of our own supporters and would be watched by a television audience way bigger than anything previously seen in the sport.

This new-fangled idea of consistency bore several harvests' worth of fruit: if we highlight another 30-month period, stretching from the 1989 Five Nations to the World Cup in 1991, the personnel changes under Geoff were minimal. Ten of those who who drew with Scotland in the first match of the 1989 championship would start in the global final against the Wallabies. There was method, there was forward planning, there was a clear direction of travel. Even if it went slightly wrong at the most sensitive point of the process (our chances of winning the great prize would probably have been enhanced by the presence of Dean Richards in the back row against Australia), we remained cohesive enough to win a second successive Grand Slam the following year, beat the All Blacks in 1993 and go very close to another Five Nations clean sweep in 1994. Even when Jack Rowell succeeded Geoff ahead of that year's summer Test series in South Africa, there was no ripping up of the team sheet. No upheaval, no great shock to the system. Jack, like Geoff, understood the concept of coherent development. During his great years at Bath, he changed his first team about as often as Margaret Thatcher changed her mind.

Leading into the 1999 World Cup, there simply wasn't the same air of stability about the national team. Clive Woodward had replaced Jack in the autumn of 1997 to become England's first full-time professional head coach – two years after most serious rugby countries – as the RFU had clearly cottoned on to the fact that it was no longer possible for someone to run a major international rugby set-up in his spare time. If it was a significant step in the organisational sense, it was also an adventurous one. Clive's reputation as a coaching nonconformist was well earned: his work with London Irish in the mid-1990s marked him out as a man determined to do things differently and, sure enough, his initial selections were so far out of left field you could barely see the point from which they originated. There were five new caps for the opening game of his tenure, against Australia at Twickenham, and he would introduce another 14 players to international rugby during the 'tour of hell' the following year. Some of his hunches were bang on the money: the likes of Matt Perry, Phil Vickery, Josh Lewsey and, of course, Wilkinson would go on to make huge contributions. But many came and went in the twinkling of an eye: Andy Long, Will Green, Spencer Brown, Steve Ravenscroft, Scott Benton, Ben Sturnham, Richard Pool-Jones, Jos Baxendell, Rob Fidler, Dave Sims, Tom Beim, Paul Sampson, Dominic Chapman, Stuart Potter. This was the 1970s revisited.

In my opinion, Clive's selection was pretty average for quite a while. In his first 28 games as coach, starting with that draw against the Wallabies in 1997 and ending with the World Cup quarter-final defeat by the Springboks in Paris almost exactly two years later, he named an unchanged side just once. I never felt he was completely sure of his next move, and was reinforced in my view by the Wilkinson incident in that last game. Jonny had been scrapping over the No. 10 shirt with the likes of Mike Catt and Paul Grayson for several months and, by the time the global tournament came around, he appeared to have won the argument. He started all three warm-up matches, accumulating 40-odd points in the process, and held his place for the opening pool game with Italy and the full-on collision with the All Blacks, who might have been beaten but for another of those virtuoso displays that Jonah Lomu saved exclusively for England. Jonny also started the quarter-final play-off game with Fiji, rattling off seven penalties and a conversion in a 45-24 victory. And then, completely contrary to all expectations – not only those of England followers, but those of some of the senior figures on the Woodward coaching staff – he was dropped for

Grayson on the eve of the meeting with South Africa. That's not how high-performance teams function. You don't chuck one of the pilots out of the Red Arrows formation on the morning of a flight when you've spent the previous six months rehearsing your routine in the finest possible detail. You don't have a bloke waking up and saying: 'Bloody hell, I'm flying today. How did that happen?'

Clive had asked people to judge him on the World Cup performance, but as some wisecrack merchant famously remarked after the 2003 triumph, he hadn't specified which World Cup. There was no change at the top after the 1999 disappointment: no change of coach, no change of captain, precious little change of any description. If it took six years and some very painful defeats for Woodward's team to fulfil its potential, the second half of that period was marked by the consistency that had so obviously been missing during the first half. Through trial and error, through a mix of accident and design, I think he found his side in the end.

Yet in the immediate aftermath, England reverted to type. Less than a year after Jonny's climactic drop goal in Sydney, things had fallen apart to such a degree that a casual observer might have wondered if Twickenham had fallen victim to some kind of apocalypse. We had surrendered our Six Nations title and been trounced by both New Zealand and Australia; Woodward had walked out; Lawrence Dallaglio, the natural successor to Martin Johnson as skipper, had announced a premature retirement from Test rugby. When England took the field in October of that year for the first game of Andy Robinson's stewardship, only four of the players who had started the World Cup final were still in place. Flabbergasting.

It is a very English trait in rugby: the inability to get our heads around succession planning. A comparison with the All Blacks is instructive. When they experience one of their flat periods, however few and far between those periods may be, there is always a sense that they know where they're going: not just tomorrow or next week or next month, but next year and the year after that. After ending a wait of almost a quarter of a century to reclaim the world title in 2011, they kept almost half their squad together for the defence in 2015. The back row that started the first of those finals also started the second; the centre pairing stayed the same; if it had not been for injury issues during the tournament, they would have stuck with the same props too. Of the 20 two-time winners of the World Cup, virtually three-quarters of them are All Blacks. The lesson to be learned is as plain as the flattened nose on a front-rower's face: it is far easier to build on an inner core than it is to fill a vacuum. And they are at it once again. The New Zealanders may have lost Conrad Smith and Ma'a Nonu and Daniel Carter from their back division; Tony Woodcock and the great Richie McCaw may have disappeared from the pack. Did anyone actually notice that the most cap-laden team in the history of the sport shed some prize assets after the 2015 triumph? The smoothness of this latest transition must have surprised even them.

One of the things common to all top sports teams is a deep-rooted system geared towards the production of the next athlete. Spanish football knows where it's going, as does German football, as do many of the more successful British Olympic sports. This clear view of the road ahead does not materialise by accident, and there is no short cut to developing a sense of direction. It is the product of a performance-behavioural programme, carefully thought through, precisely targeted and rigorously implemented, with strong foundations in terms of finance and accountability. If the English system had been functioning as it should have been when Jonny broke down after the 2003 World Cup, both the existing coaching team and the new set-up might have had a clearer idea of what to do with a youngster as talented as the Bath outside-half Olly Barkley, whose international career would never achieve lift-off, let alone hit the stratosphere.

Time and again, English rugby has enjoyed brief spells of selectorial consistency – spells that coincided with tangible success, strangely enough – before sliding back into old habits. As an economist might put it, there has been far more bust than boom. Too often, the national coach has either lacked the clarity necessary to pick a team at Test level, or found himself starved of playing options in important positions. Too often, it has been difficult not to reach the conclusion that the man in charge has picked an individual for want of a better idea – a state of mind that leads straight to the revolving door syndrome. Which is when panic takes hold, which in turn leads to decisions being made in a head spin, which in turn drags everyone in and around the squad into a whirlpool of uncertainty. It takes a strong-willed, clear-sighted character to make the most of the material available to him and to know in his own mind what next year's team will look like, which is why England finally opted to look overseas for a prime candidate and gave the job to Eddie Jones, an Australian with a proven track record of success in such diverse rugby environments as South Africa and Japan. But even a coach as hard-boiled as Eddie requires depth in his playing pool: when key players go missing through injury and there is no one good enough to step up, the best selector in the world might struggle for a proper night's sleep.

Between 2003 and 2011, in particular, such problems were much in evidence, much to our discomfort as a rugby nation. But over the last decade there has at least been an attempt to make the most of a set of advantages for which every other major union country, with the possible exception of France, would happily kill. The progress towards giving the head coach everything he needs to help him cope with the stresses of the international game has been circuitous at best and painfully slow at worst, but I believe there is reason for hope. It may just be that in the second half of this decade and the first half of the next, English rugby will make more of itself at Test level, and for longer, than any of its predecessors.

7
THE SEAT OF POWER

THE LATE AMERICAN businessman Malcolm Forbes was a master of the quotable one-liner, a gift that allows him to live on through the pages of a thousand anthologies. If some of his remarks seem just a little glib, coming from a man so bewilderingly wealthy that he could afford to collect Fabergé eggs and splash out millions of dollars on his own birthday parties, one of his comments has the ring of eternal truth about it. 'If you have a job without any aggravations,' he noted, 'you don't have a job.' Don't I know it?

At the back end of the amateur era, everything had seemed reassuringly straightforward. There had been many a bump in the road, sure, and those of us who thought about the future direction of the union game knew there would be more on the way, but for all the spats and skirmishes with Twickenham on the vexed subject of off-field commercial activity, and the occasional crisis over disrespectful comments aimed at our lords and masters, I had spent the first half of the 1990s operating within a structure I understood. All that was swept away in the space of a few transformative weeks at the mid-point of the decade, and things continued to change at an unnerving speed for the next 20 years, a period that never completely lost its Wild West feel, even when I turned my back on life in the gunslinger's saloon and took up residence behind a desk in the sheriff's office.

After the frustration of the 'Andrew Report' episode, a move to a high-profile position within the Rugby Football Union was not an obvious one and, for a good while – five years, give or take a few months – I really didn't see it coming. In the immediate aftermath of the rejection of the task-force document by the governing body's full council, I headed back to the day job at Newcastle, aware that there was plenty of hard graft ahead of me. Our victory in the Tetley's Bitter Cup final of 2001 – Tuigamala's match – had given the club a welcome lift after the relative failures of our post-title seasons and the sudden departure of Sir John Hall as paymaster in chief, but that wonderful 80 minutes on centre stage amounted to precisely that: 80 minutes. While I was more than happy with the quality of those young players who had been fast-tracked into the starting side, and ever more confident in the powers of Jonny Wilkinson (still little more than a kid himself, albeit as mature-minded a youngster as it is possible to imagine), many of our rivals in the top division were arming themselves ever more heavily with overseas know-how. Bristol had two high-quality Argentines in Agustín Pichot and Felipe Contepomi, along with a New Zealand centre as capable as Daryl Gibson; Northampton had gone south to pick up back-five forwards of serious clout in the Wallaby lock Mark Connors and the All Black flanker Andrew Blowers; Saracens had two of the most exhilarating midfield talents in the world on their books (Thomas Castaignède of France and Tim Horan of Australia), while Gloucester were flexing their muscles with the formidable Samoan back-rower Junior Paramore (running joke of the time: 'Christ, I'd hate to play against Senior Paramore') and his fellow islander, the equally forthright centre Terry Fanolua. We found ourselves running ever faster, just to stand still.

And it came home to roost in the 2002–03 campaign, when relegation and all the upheaval and pain that goes with it suddenly seemed more a probability than a possibility. Could there have been a worse moment for the earth to move beneath our feet? It's hard to think of one. Until that season, we'd been steady enough, if a little stuck in terms of our development on the field. Part of the reason was the scale of our development off the field. Dave Thompson, our chairman, felt the time was right to make something of Kingston Park as a venue worthy of hosting a top-end professional rugby team, and there were good reasons to agree with him: with its temporary seating along one side of the ground and its many other shortcomings, it wasn't much of a place when you set it against Welford Road or Kingsholm or Franklin's Gardens. Dave had long-term ambitions for the club – ambitions I shared, by and large – and the way he saw it, nothing much could be accomplished without creating a home fit for

ure bliss. I was so supercharged after my match-winning drop goal against the Wallabies at the
995 World Cup, not even Jeremy Guscott could keep up with me.

We believed we could beat the All Blacks in the semi-final a week later. Some bloke called Lomu had other ideas.

Two fellow Wasps who would join me on the barricades in Newcastle: scrum-half Steve Bates, above, and back-rower Dean Ryan.

The 1995 cup run. Celebrating victory after our semi-final victory at Leicester

'a'aiga Tuigamala, the most expensive player in the world, makes his bow on Tyneside in 1997 *above)*; Jonny Wilkinson, who would become the most famous player in the world, pretends to sten as I give him the benefit of my wisdom.

Sir John Hall, the Newcastle owner, makes a return on his investment by lifting the Premiership trophy in 1998.

Once we went international with our recruitment, we invested wisely in Mark Mayerhofler of New Zealand *(above)* and Mark Andrews of South Africa *(right)*.

Knock-out kings again. My lap of honour with centre Jamie Noon following our Twickenham
ictory over Sale in 2004 *(left)* was just a little more reserved than Epi Taione's version.

aking the long view: I join Martin Johnson and Baron's successor, John Steele, on the touchline
t Twickenham.

Twickenham's movers and shakers: Fran Cotton *(above left)*; Martyn Thomas *(above right)*; Gloucester owner Tom Walkinshaw, sitting alongside his boardroom rival Francis Baron *(below)*.

Iore men in the thick of it: Peter Wheeler, the Leicester grandee *(above left)*; Richard Smith QC, ngland's travelling barrister *(above right)*.

et the good times roll: Clive 'oodward, left, and Andy Robinson England training a year before the iumphant 2003 World Cup campaign.

Eyes on the prize: Jonny Wilkinson drops THAT goal in Sydney on World Cup final night.

Not a good look. The photographers catch me sitting behind Andy Robinson in the Twickenham stand as his England coaching career reaches crisis point.

Not a good day: Toby Flood is eaten alive by Pumas as England lose to Argentina at Twickenham in 2006.

Brian Ashton felt he could broaden England's horizons, but our first ever visit to Croke Park in Dublin was a gruesome experience.

purpose. There was no chance of us upping sticks: after the 1998 triumph, we'd experimented further with matches at Gateshead, only to find that the crowd had disappeared. We had attendances of 3,000 for Bath and 4,000 for Wasps, not even half of the audience we'd attracted just a few weeks previously when we'd been chasing the title. By the time Dave succeeded Sir John, it was a case of Kingston Park or nothing.

Thanks in no small part to my good friend John Parkinson, who was then managing director at the Falcons, we found our way through the labyrinth of a planning inquiry and embarked on the biggest capital project in the club's history. Our principal sponsors were the Northern Rock bank, and it was to them we turned. They had always been incredibly supportive, not just of us, but of sport across the north-east of the country: they'd been involved with Newcastle United and Durham cricket, as well as with the rugby. I'll never forget the crucial meeting with their board, where Dave and I outlined the vision for the club and basically said: 'If you don't lend us the money, nobody will.'

At first, I thought we were done for. 'Looking at your business plan,' said one of the executives, 'the most we should be lending you is £3.5 million.' As we were looking for £14 million, this was not an encouraging start.

But Dave stuck with it. 'We won't finish the south terrace with £3.5 million,' he said. 'We need to develop the entire west side of the ground, as well as both ends, not to mention put in new bars and corporate hospitality infrastructure.' And it was then that the people on the other side of the table looked at each other and said: 'Oh, go on then.' Neither of us could believe it. We both punched the air as we left the room.

As it turned out, this was far from the end of the story. Northern Rock went on to sign a shirt sponsorship deal for £700,000 a year, by some way the biggest in English club rugby at the time, and this enabled us to pay back the interest on the loan we couldn't really afford. Even then, the arithmetic was difficult for us. Dave eventually went back to them and negotiated a new agreement under which they would buy the ground from us, thereby wiping off the debt, and act as landlords for a nominal rent while continuing with the shirt sponsorship, which helped the business stay afloat. Ownership then transferred to Her Majesty's Government as a result of the financial crash of 2007–08, when Northern Rock was nationalised, after which Kingston Park was flogged to Northumbria University for a fraction of the original purchase price. It was only in 2015 that things turned full circle, with Semore Kurdi, the current Newcastle owner, buying back the ground from the university with help from the city council. It has been quite a saga, one way or another, although those who blame the rugby club for bringing global capitalism to its knees are surely guilty of stretching a point!

Leaving aside its effect (or otherwise) on the free-market structures of the world economy, the stadium development impacted significantly on the club, not least in minimising the amount of money available for team-building, as opposed to stand-building. The effect of this was plain to see when, on the day we unveiled our newly refurbished home to the public in early February 2003, we found ourselves six points adrift at the foot of the Premiership table with eight games left on the fixture card. It was one of those 'Oh my God' moments. We had done the right things by investing in locally grown talent and pouring money into Kingston Park, but there were no free league points on offer as a reward.

On grand opening night against Harlequins, we were up to our eyebrows in the smelly stuff – a fact brought home to me in the few seconds it took Mark Pougatch, the BBC Five Live presenter, to ask the obvious question. 'Congratulations on the fantastic new stadium,' he said to me in an interview shortly before kick-off. 'But you're going down, aren't you?'

I blustered my way through a response, something along the lines of 'the season's not over yet', but there was no hiding from the fact of the matter. I still remember that day as one of the most uncomfortable of my rugby life, as player or coach or manager or administrator. When we won 32-17, having played really well, I felt as though I'd won the lottery, cleaned up at the races *and* broken the bank at Monte Carlo, all in the space of a single evening.

That victory was the direct consequence of a change of policy on the selection front. As we slid ever deeper into trouble, we decided that while our youngsters were doing pretty much everything we could reasonably ask of them, we needed some grown-ups to help them plot a route to safety. Hence the signings of four hardened southern hemisphere professionals: the All Black centre Mark Mayerhofler, who had played against England on the 'tour of hell'; the World Cup-winning Springbok lock Mark Andrews, who had just retired from international rugby after spending the second half of his Test career performing at World XV level; and a couple of tough back-row forwards who knew their way around the track, Warren Britz and Craig Newby. They were the products of a 'needs must' recruitment drive that pretty much saved our bacon, although we would also have some valuable assistance from Bristol.

As a rugby director at the top end of the domestic game, you find yourself in the marketplace the whole time, even if you're committed to promoting from within. Why? Because unless you switch off your phone and disable your email server, you're bombarded by agents from dawn to dusk. As time goes on, you build up relationships with those you feel you can trust, and that leads to still more phone calls and emails. I saw Bart Campbell, my point man at Global Sports Management, as one of the good guys, and did some deals through him, not least in that Christmas and New Year period in 2002–03 when the team was in trouble and I agreed with Dave Thompson that it was a classic 'shit or bust' situation. There was a salary cap in place, but even though Jonny was by now earning just a little more than the £12,000 on offer back in 1997 (his wages would rise by another few bob after he dropped that goal of his in Sydney), we were in no danger of breaking it, partly because we'd been trying to manage down our losses, but mostly because the lion's share of our spending had gone into bricks and mortar. I therefore had the freedom to look abroad for players capable of staving off relegation.

Generally speaking, the words 'overseas' and 'overpayment' are closely related in any sport: the moment you go international with your recruitment, you're competing in a very different marketplace to the domestic version – a marketplace where you simply cannot hope to control the money flow. If you produce your own talent, it's relatively easy to manage the progression in terms of wages. Even with a sporting superstar like Jonny, you can save yourself a few grand by playing the loyalty card (and let's face it, few players in recent memory were more loyal than Wilkinson).

With foreign imports, it's another ball game entirely. The agent will assure you that there's a bidding war going on and that unless you cough up an extra £10,000 minimum, his man will be heading elsewhere. Is he telling the truth? Is there really another club in the race, or is he making it up? How can you be sure either way if he tells you that there's a club in France closing in on the bloke you think might just save you from a financially calamitous drop into the second division?

Sometimes, if your bond is strong enough and your luck is in, you might extract some hard information from an agent rather than a handful of white lies, but fundamentally it isn't in his interest to tell you the unvarnished truth. In effect, you're in a game of poker, holding a bad hand and saying to your club owner: 'Right, do we call the agent's bluff or throw in our cards and say we'll pay what he's asking? How far do we raise the stakes? Are we really going to tell him we're not moving on our offer and risk him phoning us back tomorrow and saying: 'Sorry, my man's off to Toulouse. Don't try to tell me I didn't warn you.'

That was where we found ourselves midway through the 2002–03 season: we knew we needed to bring in the cavalry, and the agents knew we knew. So we made the call to spend the money to get the players we required.

Those players did us proud. We might not have seen the best of Andrews across the whole of his stay with us, but when he was on his game he was every bit as good as his reputation suggested. Britz had a touch of the wild rover about him, and he would go on to play top-class rugby in France with Montpellier, but the foundations of his game, firmly laid in his native Durban, were completely reliable and exactly what the doctor ordered. Newby was pretty much the complete package as a back-rower in the grand New Zealand tradition: one look at his employment history tells you all you need to know about the range of skills he offered. As for his countryman Mayerhofler . . . Well, he was absolutely top drawer and perfect for us: in fact, I'd put him high on the list of the best overseas captures ever made by a Premiership club.

Once all four of our signings were properly integrated into the first-team squad, we had the look of winners about us again, and had the relegation scrap gone all the way to the bitter end I think we'd have survived. But it didn't go that far – not for us, anyway. Bristol were the ones left without a chair at the top table when the music stopped after Malcolm Pearce, their avuncular but less than predictable major investor, went public with plans to cut his losses and pull out of the business at season's end. The effect on squad morale was profoundly negative, the club quickly became an unmanageable mess – they were offloading players to rival teams, including the England lock Garath Archer to us, with matches still left to play – and while we were beating the likes of Leicester, Northampton and Saracens on the run-in, they went into free fall.

At this stage, a move to Twickenham was not registering on my radar. For one thing, there was no job available. England had just won a Grand Slam and were moving purposefully towards the global tournament in Australia; the club–country conflict had gone quiet, although there was often the distant sound of small-arms fire over one issue or another; and life at Newcastle was looking brighter, now that the new stadium was up and running and we'd found our way round the trapdoor leading to the second division, having looked odds-on favourites to fall straight through it only a few weeks previously. And if the 2003–04 league season turned out to be a long way short of vintage in terms of hard results, we allowed the scoreboard to get away from us on only a couple of occasions – first at Gloucester, then at Sale.

Those disappointments were as nothing when compared with the upsides: Jonny coming back to the club from World Cup duty as just about the biggest name in the sport, followed by another cup run that took us back to Twickenham, where we beat Sale in a high-scoring final, bagging four tries to their three, the last of them through Phil Dowson so close to full time that it proved decisive. Two knock-out titles in four years? I saw it as just reward for our resilience in the face of some serious threats to our very existence.

If everyone at the club was happier than they had been, so too were Northern Rock. It was they who planted the idea of going after 'another Tuigamala' – of making our rivals sit up and take notice by making an audacious play in the market. As Jonny was already ours – it would have been difficult to sign him twice – I came up with what I considered to be the next best thing.

'You want someone to light the place up? What about Rupeni?' I said. 'He'll start some fires, that's for sure.'

Rupeni Caucaunibuca, the latest in a long line of jaw-droppingly athletic Fijian wings, and perhaps the most extraordinary player ever produced by that nation, had performed a number of feats during the World Cup just past, not all of them strictly legal, but every last one of them memorable. To begin with, he had scored a 60-metre solo try in the pool match with France that had everyone questioning their own eyesight, such was the combination of pace and balance and judgement of running angle he brought to bear on proceedings. Later in the game, he caught the darkly intimidating Fabien Pelous with one of the better round-arm lefts you'll see outside of a boxing ring and was banned for two matches. Drafted straight back into the starting line-up for his country's final fling with Scotland, he scored two tries, both of them better than the blinder against France, in a performance of considerable virtuosity. He would, I thought, be just the man to give Kingston Park a touch of glamour.

It was in late 2004 that I packed my rucksack and headed for New Zealand, where Rupeni had been playing for Northland in that country's provincial championship, and for the Blues in Super Rugby. Would he actually be there when I arrived? Anyone's guess. Rupeni was a mystery to himself, let alone everyone else, and if I missed him in Auckland, where I had tentatively fixed up a meeting through his agent, I would have a problem finding him anywhere else. Rupeni came from the remote Bua province in the far north of Fiji and had moved around as a child, his father being a roving church minister. Every now and again, he would make himself scarce by heading back to the islands for a prolonged spell of catch-up with his family. People who knew the area well sometimes struggled to locate him. Me? I think I might have drawn a blank.

I had a travelling companion, at least: one J. Wilkinson, who was well on the road to recovery after his bout of appendicitis in Japan, but had yet to work his way back to full fitness. If Rupeni, or indeed any other player in the world, needed persuading to cross continents for a spell at little old Newcastle, who better to put in front of him as a kind of human carrot? Initially, we met up with the Blues team manager, Ant Strachan, an All Black scrum-half whom I had first encountered in a tough game for the British and Irish Lions against North Harbour in 1993. It turned out that in addition to his administrative duties with one of the strongest non-Test sides in the sport, he was expected to act as Caucau's minder. Whenever Rupeni went back to the islands, which was akin to disappearing off the face of the earth, Ant was the man charged with getting him back. We also made contact with Rupeni's lawyer: indeed, we stayed in his house on the west coast of the North Island, bang on the beach. Jonny was in something approaching full training by now, and for some reason he thought I was in good enough shape to train with him. He was more wrong about that than anything in his rugby life. I still feel sick at the sight of a sand dune.

At one point we met Caucau himself and gave him the hard sell, with Jonny leading the way. I'm not sure if he gave us serious consideration, or even if he had the remotest interest in what we were saying: his agent thought we were quite close at one stage during the talks, but I can't say I ever thought it would happen. In the end, he went to France for the first of two spells with Agen, neither of which were completely stress-free from the club's perspective.

It was a shame. We missed out on an amazing player; he missed out on the many and varied delights of Newcastle, which are very different to those on offer in Auckland and an entire universe away from anything to be found in Bua. There was a silver lining, though. The money we would have spent on Rupeni ended up being invested in Matthew Burke, the World Cup-winning Wallaby full-back and one of the finest players of his generation. On reflection, it probably worked out for the best. On his good days, which dawned with impressive regularity, Burke was every bit as unbelievable as Caucau. The advantage he had over the Fijian lay in his brand of unbelievability. Matt brought the precious gift of sanity to our rugby. Whatever Rupeni might have brought, it probably wouldn't have been sanity.

The ancient Chinese meant it as a curse when they told someone they hoped he would live in 'interesting times', and working with Rupeni might well have been too interesting for comfort. Could the arrangement possibly have borne fruit? His troubled history, both in France and with the Fijian national team, suggests I would have had something of a disaster on my hands, but at the time of our discussion I would have been more than happy to take my chances. At that point we'd had the Tongan player Epi Taione on the Newcastle books for four years, and as Epi could be just a little – how shall we put it? – different, I felt I understood something about the Pacific Islanders and their approach to rugby. By and large, they are as driven by a sense of community when it comes to sport as they are in most other walks of life. Get a group of them working productively together – as London Irish famously did with the Samoan 'awesome foursome' of Seilala Mapusua, Sailosi Tagicakibau, Elvis Seveali'i and George Stowers – and you can strike gold.

When Epi joined us and set about learning the ropes under Inga Tuigamala and Pat Lam, some of the rugby he produced was nothing short of magnificent, whether we played him on the wing or in the back row. When his 'family' of Pacific brethren broke up and went their separate ways, he seemed to find life significantly more challenging, and became rather difficult to control. Epi could have been anything on a union field: he was the very embodiment of that unique South Seas combination of size, pace, power, dynamism and physical explosiveness. There were occasions when he seemed to bring together the best of Inga and Pat; sadly, there were also times when he was far less than the sum of his parts. In the end, I think, he simply did not have the level of discipline needed to succeed season on season. Not that he has completely missed out on the good things in life. In recent years he has taken on the prime minister of Tonga in an election for the chairmanship of his country's rugby union and won, beating his powerful opponent in something approaching a landslide. He has also married into the family of the King of Fiji. As you do. I try to follow his progress because he provided us with some rich entertainment during his stay on Tyneside, but frankly, I struggle to keep up. Epi left us for

Sale at the end of the 2004–05 season. Our fellow northerners would go on to be crowned champions of England the following spring, thereby breaking the monopoly established by Leicester and Wasps, although their success had little to do with their Tongan recruit, who featured only sporadically in the first XV.

But the odd player departure was the very least of my problems. Slowly but surely I became aware of a sense of unease, although it was difficult to identify its source. We had reached the quarter-finals of the Heineken Cup in 2005 and, although we caught a very good Stade Français side at the wrong moment and were smashed 48-8 at Parc des Princes in Paris, there were a substantial number of Geordies in the 45,000 crowd and they revelled in their experience of the European high life. We also felt we had a squad with unlimited potential for growth. Jonny was still in his mid-20s and he had an ocean of promising back-line talent around him, from Lee Dickson and Hall Charlton at scrum-half to David Walder, Toby Flood, Jamie Noon, Mathew Tait and Anthony Elliott further out, their youthful extravagances of spirit balanced by the know-how of Mayerhofler and Burke. We even had some decent prospects up front in the contrasting shapes of David Wilson at tight-head prop, Geoff Parling at lock and the no-nonsense Dowson in the back row. All three would go on to do their bit for England – and, in Parling's case, for the Lions.

Together with Dave Thompson, who in his heart of hearts

was far happier putting his money into youth development than investing it in the overseas transfer market, I indulged in a little war-gaming: we would map out our squad with an eye on the future, predicting its strengths and weaknesses four or five years down the road. But Dave was showing signs of frustration with both the financial and operational sides of the venture. We were still finding it difficult to make sense of the balance sheet, and our progress on the field was slow. We were treading water, basically, and I suspected then that Dave's heartfelt commitment was beginning to weaken. The proof of it would emerge soon enough.

Not long after my departure, he began to let people go. Walder said his goodbyes around the time of my own farewell; Dickson, Flood and Tait were out by the middle of 2008; Parling, Dowson and Jonny – yes, even him – left in 2009. Not that I would pin so much as a molecule of blame on any of them, least of all Jonny. He'd been incredibly loyal for a very long time and, while the club had shown loyalty to him too, particularly during the years of injury, he was perfectly entitled to do the right thing by himself and for himself by moving to Toulon and helping that once-great French club become greater than it had ever been. If I'm honest, I don't think Newcastle were ever quite able to give him the platform later in his career that he had at the start, when he was surrounded by world-class players; he slowly outgrew the club because we couldn't grow with him, despite having such a good crop from the academy. However, I do still wonder where that team would have gone had we held it in one piece and supplemented it with a couple of gnarled forwards in influential positions. Dave and I spent enough time thinking about it. It is a sadness of mine that we'll never know what they would have achieved had we found a way of keeping them together.

Yet it was not only Dave who was wrestling with his thoughts. There was some push and pull at work against a background of widespread turbulence in the English game, and it had its effect on my thinking too. Twickenham was going through another of its bloodthirsty spells: on the so-called 'day of the long knives' in April 2006, nine of those with hands-on roles in the England set-up were sacked, including high-profile coaches in Phil Larder, Dave Alred and Joe Lydon, while the performance director Chris Spice resigned. Andy Robinson, who had been promoted to the head coaching role when Clive Woodward walked out of HQ in a blaze of protest and recrimination some 18 months previously, was still in place, much to the heavily theatricalised 'disgust' of those in the press who accused him of saving his own skin by sacrificing everyone else – a pretty cheap shot, even by the worst standards of Fleet Street. The top end of the game was in a mess, again, and it was clear that Francis Baron, sitting behind his chief executive's desk, was devising a new approach.

The broad detail of this quickly emerged: the union would appoint a director of elite rugby, supposedly with unprecedented powers over the running of the Test side and associated matters. What was more, they went to the market in the most public of ways by advertising the position and appointing a team of head-hunters to find the right man. It was not long before senior rugby journalists up and down the land, most of them strongly briefed by their contacts inside Twickenham, set about lining up the runners and riders for the big race. Clive was among them, inevitably, even though he had so frequently mocked the idea of a head coach reporting to a rugby director. (When he was doing the job, he clearly had very little time for Chris Spice, who, tellingly, did not merit a single mention in Clive's 2004 book, *Winning!*. Chris was very much an ideas man and almost certainly had more to offer English rugby than he was allowed to show.) Other names being bandied around were Nick Mallett, the former Springbok coach who had led Stade Français to a couple of French club titles; Eddie Jones, then in the middle of a consultancy role with Saracens; and Ian McGeechan, of Scotland and Lions fame. I was also on the list.

In the penultimate game of the league season, at the back end of April, we were playing Worcester at Sixways. I remember being approached by a group of reporters who asked me if I was 'throwing my hat in the ring', as the saying goes. I wasn't throwing anything anywhere at that point and I told them so. 'Look,' I said, 'I haven't finished my work at Newcastle. We have a new stadium and a new young team. There's plenty still to do.' Which was a genuine response. A big part of me was still excited by the shape of the team: a week later, when we ran down the campaign curtain with a home match against Leeds, our entire match-day squad was England qualified. Some saw it as a stunt, and there was certainly an element of us wanting to make a point, but we won 54-19 and racked up eight tries. Point made, I thought.

There was, however, something I did not share with the journalists at Worcester. I had indeed been approached by the head-hunters, who were putting together a shortlist and had made it clear that they wanted my name on it. We talked it through, then talked it through a second time. Usually, I have a fairly clear mind when it comes to this kind of thing. Here, I was indecisive. There was a part of me open to temptation, a part of me itching for a change – but equally, Newcastle still had a hold on me. I'd given a lot of myself to the job on Tyneside and, while I was uncertain about some important aspects of the club's future, I was far from convinced that it didn't have one. After weeks of thought, I reached the conclusion that the RFU job would be the wrong move at the wrong time and told the head-hunters that I wouldn't be a candidate.

And then I changed my mind. There were a number of triggers, over and above any unease at what I feared was Dave Thompson's diminishing enthusiasm for life at the top end of professional club rugby. The main one was the deteriorating relationship between the Premiership clubs and the governing body – yes, that old chestnut. You might think that anyone in full control of his faculties would have seen this as an insurmountable barrier rather than a trigger; that after the numbing experience of the task-force episode in 2000, anyone associated with that ill-fated project would have reacted to a resumption of political squabbling by running in the opposite direction at Caucaunibuca-like speed. But it was precisely because I was so completely fed up with the politics that I decided to reinvolve myself. It was 1999–2000 revisited. Francis Baron, still chief executive of the union, and Tom Walkinshaw, still the lead negotiator for the clubs, had resumed their mutual mud-slinging, rock-chucking approach to rugby affairs; there was a fearful row underway about Francis's decision to stage a fourth autumn Test, outside the agreed international window, against the touring All Blacks (he wanted to open Twickenham's redeveloped South Stand with a grand event; the clubs said they would not release their players for what they saw as a vanity project); and England's results on the field were way below the expectations of the union hierarchy, who apparently saw no link between their own inadequacies and the bad numbers on the scoreboard. I felt I could at least make a contribution in moving the warring parties towards some kind of reconciliation.

I understood exactly how the clubs were thinking and why, but I also had the wellbeing of the national team at heart. No player who ever pulled on the white shirt wants to see England struggle, and I was no different. A few days after rejecting the head-hunters with a 'thanks, but no thanks' message, I phoned them back and said: 'I don't know how far down the road you are with this process, and it may be too late, but I'd like to put my name forward.'

As I understood it, and still understand it to this day, I was joined on the shortlist by Clive Woodward and Ian McGeechan. I'm not sure how serious Geech was about taking on the role: he'd been absolutely brilliant with me as my Lions coach in Australia in 1989; we'd worked closely together in New Zealand four years later and I felt I knew him well enough, but there are times when he can be a little opaque and move in mysterious ways, his wonders to perform. Clive was certainly up for the job, and he had the backing of Martyn Thomas, who had been elected to the RFU chairmanship the previous year with heavy grassroots support, and who would become one of the most divisive figures ever to find his way into high office at Twickenham. When I hopped on a train down to London to talk about the responsibilities carried by the brand-new role under creation, I was not at all sure that when push came to shove, my candidacy would be successful.

I knew I had no interest in actually coaching England: I'd done a bit of tracksuit stuff at Newcastle but didn't really enjoy the hands-on side of it as much as I did the recruitment, the development of the business model, the work on the new stadium and the dozens of other aspects of running a rugby club. My view was that the RFU's elite director should oversee the rugby side of the English game in such a way that the other senior figures could concentrate on their own jobs rather than waste time entangling themselves in everyone else's. At the time, it seemed that Andy Robinson was being dragged more and more into off-field issues, while Francis was too involved with rugby-specific problems. The England coaching job is difficult enough without endless political wrangling over player release, money, and all the rest of the crap that attaches itself to the management of a major professional sport at the highest level, yet Andy was fire-fighting the suits and blazers for most of his working week. If you're not in a position to do your job properly, guess what? You don't do your job properly. I knew I didn't have all the answers, but I was clear in my view that there had to be a better way of running the show: that for all the compromises that would have to be made, there was surely a deal to be reached that would allow the England team to prosper without disenfranchising the clubs. I had a loyalty to club and country. To me, it didn't have to be one or the other.

There was a four-man interview panel. Francis was there as chief executive, as was Martyn Thomas as chairman, accompanied by two former England captains in Bill Beaumont and John Spencer, both of whom had been extremely active at RFU committee level for as long as most of us could remember and could pick a route blindfold across the minefield of committee-room Twickenham.

It seemed to me that Clive Woodward had a surge of support from his mouthpieces in the press, and he clearly had Thomas on his side. There again, I was confident of having John in my corner, along with Francis Baron, if only because his falling-out with Clive in 2004 had been so spectacular. My subsequent take on events was that if Bill just about favoured my opponent, he was nowhere near ready to go to war over it. Even had the panel been split down the middle, Francis would have had the final say. Either way, I was the one who ended up being offered the job, and as I'd already had Dave Thompson's blessing, ('If you want to go for it, I won't stand in your way,' he'd told me.) I accepted. It was a wrench, leaving Newcastle after everything I'd been through there, but in reflecting on those 11 turbulent years, I realised I'd become a little stale. It might also have been the case that Dave shared that view and was not wholly against the idea of freshening things up with a new rugby director. He never said as much to me, but I'd have understood it if he had.

What did I find when I arrived at Twickenham to start my new job? Quite quickly, I found myself missing Tyneside. Whatever the complications of life at Kingston Park, they were as nothing when set against the challenging nature of the role I'd just taken on. The England set-up was in a bad place and would quickly deteriorate from there, while the club–country relationship was back to its dysfunctional worst, to the extent that there had just been a High Court case over the additional autumn fixture with the All Blacks. To make matters more depressing still, I knew we were miles off the pace in terms of representative age-group rugby – the third of the key elements crying out to be addressed, and by no means the least of them. There was no remaining trace of the World Cup-winning euphoria that had held things together in the first half of 2004, and the existing agreement between the Premiership teams and the governing body would soon be up for renegotiation. That little task came under the heading of 'over to you, elite director' and there was no escaping the magnitude of it. Without a meaningful deal, there could be no progress. English rugby would be knackered. Completely knackered.

It was obvious to me that we needed an English solution to an English problem – a problem necessarily different to those in Wales or France or South Africa, for the very good reason that we weren't Wales or France or South Africa – and that the only possible answer was to strike a deal that worked for club *and* country. I thought of the 2003 World Cup victory, and still think of it now, as a blip. Not in terms of the performance of the team: it was a genuinely great team, the best we've ever produced, full of outstanding players who happened to be in the same set-up and who were strong enough to deliver under paralysing pressure. It was a blip in terms of the fundamentals of our game. It was the product of a moment in time, not the product of a programme geared towards consistent success. We needed a new settlement that had nothing to do with the old thinking. Central contracts? They were gone for ever. Franchises? They were gone too. Owners with money and willpower and the determination to protect the club game from RFU domination? Still here. Power politics? Still here. It was going to be a hell of a job, but at least I understood what I was trying to achieve.

I needed a starting point with the clubs. I knew I could speak their language – I'd been a part of the cause for long enough to understand their aims and motivations – but such was their lack of trust in the union, it was not easy to find a patch of common ground on which we could all stand. Ironically enough, we were able to coalesce, however tentatively, around one or two of the main conclusions put forward in a report titled 'The Way Forward' – a paper compiled by the multinational management consulting firm L.E.K., to whom Twickenham's grandees paid a pretty penny. In one sense, that cash could have been put to better use: L.E.K.'s analysis of the professional game, carried out before my arrival at Twickenham, was effectively a second version of the Andrew Report. Talk about groundhog day. When I read through it, I recognised an awful lot of arguments concerning game structure, academies, player welfare and the distribution of monies. But L.E.K. went much further in attempting to breathe new life into the central contract argument. Looking back on it now, I have a suspicion that in commissioning this piece of work the RFU was playing a game of double bluff, of black propaganda. They were effectively saying to the clubs: 'There's some stuff in this paper that you won't like, so if you don't play ball with us now, we'll chuck everything up in the air and go back to square one.'

Yet there was enough in the document to form the basis of a discussion, so the talks were soon underway. I led for the union, with Francis alongside me and Martyn Thomas making the occasional appearance. (It seemed to me that he wasn't much interested in a deal. He believed that if a proper, mutually beneficial agreement remained out of reach, the clubs would be starved into submission and would therefore have no choice but to give the RFU everything it wanted.)

On the club side, Tom Walkinshaw was the lead negotiator, supported by Mark McCafferty, the chief executive of Premier Rugby Ltd, the clubs' umbrella organisation, and Peter Wheeler, one of the finest of all England tight forwards and a powerful influence at Leicester, the biggest club in the country. Other owner-investors dipped in and out of the talks – Nigel Wray of Saracens, Keith Barwell of Northampton – but the main work was done by the ever-presents.

Peter was a particularly important figure as far as I was concerned. Much as I liked Francis, I don't think he ever quite knew what he was trying to achieve in this arena: he couldn't quite get his arms round the issues somehow. It was only when he was trading punches with Tom that he seemed remotely sure of his position. My relationship with Peter was far more constructive. He was my go-to rugby man; someone to whom I could say: 'Let's go and talk things through in a quiet corner somewhere and let them get on with the alpha-male stuff. If we give a bit here and you give a bit there, we can make some progress.' We understood each other. We'd both played for England and were passionate about the fortunes of the national team. At the same time, we both knew something of the realities of club rugby in times of financial hardship and political brinksmanship. We had a connection.

As the hard bargaining unfolded over many weeks, a deal began to take shape. There were three elements. At the top end, there was the idea of an Elite Player Squad (EPS), based around the things our international players needed if they weren't to take on the best Test teams in the world with one arm tied behind their backs – or, as happened more than once in my early days in the job, both arms wrenched backwards and repositioned painfully between the shoulder blades. They needed proper rest, proper medical management and proper preparation. None of those things was currently in place. The middle segment was an England Qualified Player (EQP) arrangement, under which clubs would be incentivised to pick people in their first teams who were eligible for representative rugby. This was a sensitive issue: we could not make the mistake of introducing quotas because these ran contrary to European law. Had we insisted that the clubs play 15 home-qualified players on a Saturday afternoon, we'd have been in court before lunch on the Sunday. But a voluntary arrangement, made suitably attractive with a little financial encouragement, could be made to work, especially if the third plank

– the establishment of an academy structure capable of producing the right standard of player in the right physical condition – was also agreed.

This was of special interest to me, having overseen the emergence of so many outstanding youngsters on Tyneside. No rugby director at a Premiership club would fast-track a player into his senior team simply because he wanted to do the right thing for England. How did I know this? Because I'd been in that position myself. This kind of selection policy, patriotic though it might be, could easily leave someone vulnerable to sacking. The only way the bright young things get to play is if they're good enough and strong enough to play. If those people aren't available at a club in need of results, there is only one thing the DoR can do: go to the market and find himself an Andrews, a Mayerhofler, a Newby or a Britz. If, on the other hand, the academy manager comes to you and says, 'There's no point throwing money at elderly New Zealanders and ancient Aussies because little Jonny and little Toby and little Mathew can stand the heat', and turns out to be right, the DoR will spend the rest of the season asking him if he has any more where they came from.

The bolstering of the academy system was a crucial part of the 2006–07 negotiations: so crucial, in fact, that its positive effect on the England side continues to grow more evident by the day. We now have a programme capable of churning out players like Billy Vunipola, Mako Vunipola, Joe Launchbury, Courtney Lawes, Maro Itoje, Joe Marler, Jamie George,

George Ford, Owen Farrell, Jonathan Joseph, Anthony Watson, Jack Nowell, Jonny May, Elliot Daly. That wasn't the case a decade or more ago. At that stage, the kids in England were capable of being good enough, but no more than capable. The programme was not right, and its flaws left them exposed to some horrible beatings at international age-group level. I saw it with my own eyes when the Under-19s went to Ravenhill in Belfast for a game against their New Zealand counterparts and were absolutely pummelled. We had some good prospects on the field that day – Alex Goode and Alex Corbisiero among others – but the opposition had Israel Dagg, Sam Whitelock and Ryan Crotty, all of whom seemed to be in a different kind of physical shape entirely. It was men against boys; the image still lingering in my mind is of a bunch of Jonah Lomus against normal-sized humans – and it bordered on the dangerous. England had never won an age-group title and now I knew why.

We can all be susceptible to feeling misty-eyed when we sit back in our armchairs and indulge ourselves in the rugby of our dreams. Those who have been around the game long enough talk endlessly of Barry John and Phil Bennett and Gerald Davies; of Andy Irvine and Serge Blanco and Didier Codorniou and David Campese; of a whole bunch of twinkle-toed, fleet-footed rugby adventurers whose attacking panache propelled the sport into the realm of beauty. But the inescapable fact that lies at the heart of rugby has precious little to do with the beautiful: to misquote the poet Keats, 'brutal is truth and truth brutal'. To play the game in its higher forms, you have to be physically equipped to do so. Otherwise, you're an irrelevance, no matter how skilful you may be or how clever you are in your understanding of the dynamics of the contest. I've never had much of a liking for the term 'gym monkey', but it was clear to me when I moved to the RFU that there was only one way not to be smashed by the major southern hemisphere nations at age-group level, and that was to improve our conditioning to the point where our opponents would quickly realise we were not there to be pushed around.

The people involved in the 2006–07 negotiations might not have been the first to recognise that the New Zealanders and the rest had left us trailing in this area, but there can be no denying the impact of the agreement we eventually signed. Indeed, the improvement in age-group returns kicked in almost immediately. Nigel Redman, a stalwart of the relentlessly successful Bath side who were masters of all they surveyed at the back end of the amateur era, took the England Under-20 team to a World Cup final in Wales in 2008, thereby breaking new ground. On the way through, they beat an Australian side boasting David Pocock, Quade Cooper and Will Genia, no less, and while the New Zealanders were too strong for them at the death (and it beggars belief, but the brilliant All Black scrum-half Aaron Smith was nothing more than bench material for them in that tournament), we had put down a marker. The Kiwis beat us again a year later (Aaron Cruden and Zac Guildford were the star performers in that silver-ferned side) and they were still a little too good in the 2011 final, even though that England team fielded Owen Farrell, George Ford, Elliot Daly, Mako Vunipola and Joe Launchbury and may have been the best Under-20 team we've ever put together. But we won the title in 2013 (Anthony Watson, Jack Nowell, Henry Slade, Jack Clifford) and retained the title 12 months later under Maro Itoje. When the tournament went to Manchester in 2016, we walked it, sticking 45 points on a high-performing Irish team in the final. It is worth pointing out that our triumph, the third in four years, could not be ascribed to home advantage – the 2013 competition had been played in Italy, the 2014 version in New Zealand – and had still less to do with the weather in the north-west. It rains in other parts of the world, including Dublin. And in Auckland, come to that.

I'm not sure any of us envisaged quite this level of achievement when we were hammering things out around the table in the autumn and winter of 2006, but I was determined that we should at least put the right conditions in place as part of any settlement. In essence, I was trying to create a system of quasi-central control driven by, and delivered through, market forces. 'We'll give you the money you need,' I was saying to the clubs, 'but every penny of it must be conditional. If you want the cash, you'll have to deliver in areas we think are important for the England set-up.'

The money on the table was substantial – £100 million over eight years, which was pretty much double the amount the clubs had been receiving from the governing body until that point – and it was divided into three pots. If the club academies were properly managed and ticked all the boxes in terms of delivery, the RFU would pay 50 per cent of the running costs. If England-qualified player targets were hit, another pile of cash would be transferred into the clubs' accounts. If they played ball over England player release and all the rest of it, the balance would follow. It made complete sense to me: the academies would drive the EQP programme, which in turn would drive England's success at full international level. There it was: an English solution to an English problem.

These things are never entirely straightforward, though: talks between different people with different views on life, people who do not necessarily share your own motivations and have alternative endgames in mind, are by their very nature a complex business. I could never quite understand Clive's antipathy towards the general theory of negotiated settlement: it used to annoy me no end when he used a tame newspaper to tell us that 'there can be no compromise if we want to be the best'. It was nonsense on a number of levels, the most blatant of which was the fundamental ignorance of the fact that the rules of the game had flipped in August 1995, when the RFU had declared their one-year moratorium on professionalism and thrown the sport open to forces they should have seen coming but didn't, and should have understood but couldn't. If Clive had been a big figure in New

Zealand rugby, he could have been as uncompromising as Attila the Hun and got away with it. Being a big figure in English rugby demanded just a little more in the way of subtlety and nuance. No matter what the more unreconstructed members of the hierarchy might have felt about things, it simply was not possible for the union to keep the clubs on a choke chain: even in the darkest days of 1999–2000, when the club game was on its knees, the owner-investors fought tooth and nail to retain the players' loyalty and succeeded, despite having nothing much to offer them. Having found a way through that trauma, they were unlikely to backtrack now.

Bringing together a bunch of personalities as diverse as Tom Walkinshaw, Mark McCafferty, Peter Wheeler, Francis Baron and Martyn Thomas and holding them close enough to reach a deal was not the easiest problem to solve, but as the talks continued there was an increasing amount of goodwill from most of the participants. There were a number of occasions when Tom, whose business interests were primarily in Australia by this point, would fly from Sydney to Heathrow for a meeting at the Crowne Plaza airport hotel, negotiate all day and into the evening, get on a plane and fly back again. I had the feeling with him that his enthusiasm for any kind of breakaway had dissipated, that he no longer had the energy – maybe not even the interest – to launch another uprising against the union. The Bernie Ecclestone fixation had gone. His driver now was the money from the central pot: how much were we prepared to chuck in the direction of the clubs, and what did we want in return? I knew I could depend on Peter to play a sensible hand, while Mark was always capable of making constructive contributions when we reached an impasse.

In many ways, the more intractable problems were with my own side. I suspect Francis rather enjoyed his games of bluff and double-bluff and, with the L.E.K. report in his back pocket, he was able to indulge himself in brinksmanship. As for Martyn, he seemed keener still to take things to the edge, if not over it. Even after England's horrible defeat at the hands of the Irish at Croke Park in the 2007 Six Nations, the scale of which concentrated most of our minds and persuaded virtually everyone involved in the talks that there was no more time to be wasted in securing an agreement that would allow the sport to move forward in our country, the RFU chairman seemed uninterested in compromise. He had been elected on an anti-professional club platform, I think he still felt those clubs could be broken and remained utterly convinced that with the smack of firm leadership, central contracts remained within the grasp of the union.

Fortunately for the rest of us, and for the game as a whole, there were enough people in high places who believed he was on the wrong side of history. Together with the RFU's finance director, Nick Eastwood, who was an extremely positive force in circumstances where negativity could easily have taken hold, Francis increasingly threw his weight behind an agreement. At Twickenham board level, Bill Beaumont and John Spencer also made it known that they rather fancied the idea of peace. I think Martyn would happily have sparked further discord if he could have found a way of doing so, but in the end, too many of the big players took the wise decision to back away from it. If it felt like Martyn wanted to trash the thing, the others were saying: 'Hang on a moment, what will be left after the trashing, apart from trash?' Martyn could not paint a convincing picture of life without a compromise with the clubs, so compromise it had to be.

It was not until the 2007 World Cup in France that we finally sent our puff of white smoke into the Twickenham sky. The final stages of the talks were extremely tiring – not least because I was spending half my week on the far side of the water trying to work out what the hell was going on with an England campaign that, in the early stages of the competition, was just about as grim as anyone could have imagined. Backwards and forwards across the English Channel I went, balancing these two major parts of my new role as best I could. But when the eight-year settlement was at last accepted, the feeling of satisfaction was intense. There was still some resistance among the backwoodsmen on the RFU council – the clubs were receiving a significant uplift in financial terms, after all – and one or two of the Premiership contingent were not wholly convinced by the terms of the agreement, fearing too much had been conceded to the governing body. But the political crust was not as thin as it had been half a dozen years previously and, in the end, there was as broad an acceptance of the final document as I could realistically have wished to see. If leaps of faith had been taken on both sides, few of the people who really mattered believed in the alternative.

The part of the settlement that encouraged me more than anything else was the acceptance on both sides that if we were going to make serious progress in a world game that had put itself on a firmer, faster track than the one we were on, we had to build from the bottom up. The notion dear to many of the union's traditionalists that the England team was the only thing that mattered, and that there would be trickle-down benefits to those standing below if only we won our autumn internationals and Six Nations matches, had finally been put to one side. If you're building a house, you don't start with the roof. That much should be obvious to anyone who ever received a Lego set as a Christmas present, but there were people in influential positions at Twickenham in the mid-2000s who saw it very differently. True, we built from the top down at Newcastle by buying a team of outsiders, but we also laid strong foundations at the same time, by being among the first clubs in England to put an academy in place and going into the northern universities in search of talent. We put ourselves ahead of the game. Even if you disregard Jonny Wilkinson's emergence on the grounds that he was a freak of nature, the development of Walder, Noon, Tait, Flood, Tom May, Michael Stephenson and all the others points to a successful system born of enlightened thinking.

Even more relevantly, the England team is not a club team. You can't simply go to the market and buy yourself a tight-head prop if a hole suddenly opens up in your front row at the start of a Six Nations championship or a World Cup tournament. One of the first conversations I had with the RFU hierarchy when I arrived at Twickenham in the early autumn of 2006 concerned the Test No. 3 shirt. 'What are we doing to help David Wilson?' I inquired, referring to the bright young scrummager we'd been bringing on at Kingston Park.

'Why do you ask?'

'Because he's going to play for England.'

'What makes you think that?'

'Well, he weighs 19 stone, he plays on the tight head, and he's the only one we have who is not of pensionable age. Phil Vickery, Julian White, Duncan Bell? How much longer are they going to be around? Who else is going to anchor our scrum? It's not going to be me.' There were a lot of blank expressions on a lot of faces that day. A decade on, such conversations no longer need to be had.

Did my move from Tyneside to Twickenham make me a classic example of the poacher turned gamekeeper? It was an easy accusation to throw and there were a good number of people in English rugby who could not resist the temptation. Yes, during my time with Newcastle, I was pretty miffed, more than once, when Jonny jumped in his car and headed south for an England training session or disappeared on tour for weeks on end. I'm sure Clive, during his spell as national coach, was equally miffed that Jonny didn't spend more of his time with England. And now, with my RFU badge on my business blazer, I was singing a different tune. But I took on the role in the deep-seated belief that I could bring club and country together for the benefit of the players and the coaches, and ultimately for the English game as a whole. Does that equate to a betrayal of principle? Not in my book.

I knew how the directors of rugby at the Premiership clubs felt when they were forced to play important games without their best players – games that could, if they were lost, lead to relegation and, pretty much as night follows day, a P45 in the morning post. I'd been there, and I'd had first-hand experience of that sick feeling in the pit of the stomach. Yet I also knew how things looked from the other side of the divide – the side on which I had put myself by throwing in my lot with Twickenham. There was a fair bit of soul-searching involved in making that decision but, in the end, I felt there was a mission to be undertaken by someone and that I had the relevant credentials. And here's why.

Before those pioneering days with Newcastle, there was nothing I had loved more than turning out for Wasps on a Saturday afternoon. The same went for Jeremy Guscott and Richard Hill and Gareth 'Coochie' Chilcott down at Bath; and for Will Carling and Brian Moore and Jason Leonard at Harlequins; and for Dean Richards and Martin Johnson at

Leicester. But there was something we all loved every bit as much, and that was playing for England: there was not a murmur of disagreement when, on gathering together for what masqueraded as a training camp in those days, Will, as captain, said: 'Jerry, Cooch, Brian, Deano . . . okay, we've all been battering the hell out of each other and had the time of our lives, but we're here now and we want to be as successful in this environment as we are in that one.' And for much of that period, there was a system in place that made success possible – that made England rugby an easy thing to love.

Was that still the case in 2006? I didn't think it was. I'd thrown my weight behind Sir John Hall and the Newcastle project because he was prepared to create something of value in a rugby world where all the old certainties had disappeared, virtually overnight. But there was nothing of value being created at international level. The thing was falling apart, we were on the road to self-destruction and that couldn't be right.

If you're lucky enough to be an international-standard rugby player, there is nothing worse than to find yourself in a Test team that is barely able to compete because the support structure has broken down. I'd experienced it myself in the mid-1980s and knew what a horrible place it could be. You've reached the pinnacle of your chosen team sport, you have every right to feel good about yourself; yet here you are, walking out of the tunnel on the world stage – at Twickenham or in Dublin or Paris or Johannesburg – in the certain knowledge that you're about to get your arse kicked. That's bad. And to exacerbate it all, you know that it's you and your colleagues who will get the blame. The coach might also find himself in trouble, but there are only 15 people out there who are about to be licked in public, and you're one of them. It's the grimmest feeling, to the extent that you don't want to be there. There isn't a player in existence who thinks to himself: 'I'm going to get a right old spanking this afternoon, but I'll enjoy it all the same.'

I'd had my fill of the downside early in my own international career and seen the same thing affect Jonny when he first broke into the England side. So with the euphoria of the World Cup victory nothing more than a fast-fading memory, replaced by a feeling of dread whenever we ran into international opponents who knew what they were about, I said to myself: 'Come on, there has to be another way. Let's get involved.'

Whether people believed in the purity of my motives was, when all was said and done, neither here nor there. I had found it difficult to sway opinion as a player: a little like the rival factions in 1980s track and field, when you were either a Sebastian Coe-ite or a Steve Ovett-ist, there was an unbridgeable gap between those who thought I should play at outside-half for England and those who felt the shirt belonged to Stuart Barnes of Bath. Now that I had swapped my modest office at Kingston Park for high office at Twickenham, I gave little for my chances of winning support from those who believed someone else should be running the England show.

But that was fine. I was confident of my own position regarding the issues that needed to be sorted, and equally confident that I had the ability to make a difference.

It has never been my way to shout about what I am doing or why I'm doing it, and I'm not the sort to cultivate the press, to get a national newspaper to act as my cheerleader. To me, delivery is all that matters, however long it might take. I'm not a flaky sort; I don't give up on something when the going gets tough. It's far more important to me to form a clear idea of where I'm trying to reach and hold to that idea regardless. My idea in 2006 was to bring clubs and country together in as fruitful a way as the circumstances allowed, and I feel that aim was achieved. Maybe I wouldn't have been appointed to the job in the first place if Clive and Francis had not fallen out so comprehensively in the months after the World Cup victory, but they did. As a result, the RFU gave the job to someone who could at least look the club owners square in the eye and say: 'I've been fighting the good fight alongside you for 11 years. You trusted me then, and you can trust me now.'

8
COACHES COME, COACHES GO

TWICKENHAM, MURRAYFIELD, STADE de France, Croke Park. . . and the Burleigh Court complex at Loughborough University. Three rugby cathedrals, one theatre of Gaelic sporting dreams and an unremarkable residential conference centre tucked away in the corner of a college campus in Leicestershire, a drop goal's distance from the M1. I have nothing against any of these places – I had more than my fair share of joyous moments on the old cabbage patch in south-west London, emerged from my international business in Edinburgh just about in credit, and never had any direct dealings with the other three locations – but during my first 18 months on the RFU payroll, each and every one of them felt like hell. By comparison, Kingston Park on a wet day, with the wind blowing straight off the Tyne and the scoreboard spinning in the wrong direction, was my idea of heaven.

They were fraught times. I expected life to be challenging when I began work as England's director of elite rugby, and of course had things been working perfectly, the position would not have been created in the first place; but I was not wholly prepared for the magnitude of the issues spilling out of my in-tray. Only weeks before there had been a High Court wrangle between my new employers and representatives of my old ones, who had objected to the addition of a fourth international match in the autumn of 2006, a move instigated in characteristically cavalier fashion by the RFU chief executive Francis Baron, who had just presided over major construction work that completed the Twickenham 'bowl' and thought that the All Blacks would be just the people to put on a grand opening show for the paying public. Andy Robinson, who had succeeded Clive Woodward as national coach two years previously, and was under a good deal of pressure after a couple of below-the-fold finishes in the Six Nations and two painful thrashings in Australia, did not see it quite the same way.

The first 'official' Test of the autumn was a must-win game against Argentina. The last team Andy needed to see inserted into the fixture list at the last minute were New Zealand, especially as it left him with only a six-day turnaround before the most critical contest of his coaching career. By walking into Twickenham when I did, I was certain to be caught in the crossfire. Even before I could turn my thoughts to the club–country negotiations, the age-group crisis and the task of defending our status as reigning world champions in France in precisely a year, it was blindingly obvious that there would be bad moments ahead.

The England team's performance on the field, together with its development off the field, was now my responsibility. That's what it said in the job description, more or less, and that's how it was reported in the media, more or less. The key words here are 'more', 'or' and 'less'. The print on my contract might have been in black and white but, in reality, the demarcation lines were blurred. This was not Newcastle: there was no blank sheet of paper, no room for the pioneering spirit, no sense that I could make this up as I went along with the full support of the people who were paying my wages. This was England, the shop window of the sport. There were things in place that would prove very difficult to change; there were men in powerful positions who were not in the habit of letting that power go to waste. In retrospect, I should have asked more questions, and demanded more answers, during the recruitment process. I should never have put myself in a situation where the most public part of my job, not to say the least forgiving, was the one made doubly complicated by a lack of clarity.

By the time I started on the first day of September 2006, Andy was already in a world of trouble. No one in English rugby doubted for a second that he was an outstanding forwards coach: his work with the World Cup-winning side under Clive merely confirmed what everyone had known since his early tracksuit days at Bath. But he had saddled himself with a terribly difficult task. The structural problems around the maximisation of England's potential at Test level

– problems that had to a large extent prompted Clive's headline-grabbing resignation in 2004 – were still there. The relationship between the clubs and the governing body was at a six-year low; results were heading in the same southerly direction; there had just been a round of painful sackings that had clearly impacted on Andy's highly developed sense of honour; and there was an increasing amount of chatter around his perceived failings as a selector.

It was against this background that I hopped in the car and drove to Burleigh Court, where the England players had congregated for an early season squad session geared towards preparation for the All Blacks match. Under the existing agreement, Andy was entitled to pull in his Test candidates for a two-day gathering in September and another get-together in October. It wasn't much – no England coach would have the luxury of full, blocked-off access to the players for a significant amount of time in the build-up to an international series until the 2007 accord had been ratified – but, as there was nothing else available, he had to make do and mend. As I wandered down to the training field, it seemed to me that there were rather more players in attendance than the 40 I had expected. In fact, there were 55: a few of them gambolling around on the pitch, and rather more – including most of the senior international hands – looking on uneasily from the touchline. Over a cup of tea, Andy told me that most of the players he wanted to see train were crocked to a greater or lesser degree: they'd played Premiership matches for their clubs that weekend and were either genuinely unable to take an active part in the session, or had been told by their clubs that they were too injured to pull on a pair of boots. So he'd called in a few extras to make up the numbers, some of whom might not have been household names in their own homes.

After 11 years of club management on Tyneside, I knew the score. Premiership coaches always loathed those early autumn England call-ups: you go through a hard pre-season, play your warm-up matches, throttle up through the first two or three fixtures of the league campaign . . . and then find you cannot prepare properly for the next game because your front-line personnel have gone missing for 48 hours or more.

Now, looking at it from an RFU angle, I saw the same problem differently. There was zero value in that session at Loughborough: a majority of the players pencilled in for Test duty in six weeks' time were standing around doing nothing and, if truth be told, they would rather have been somewhere else. Andy had called them in, along with all those who didn't stand a cat's hope of facing the All Blacks, because the training days were on the calendar and had to be used. Talk about a waste of time, money and energy. We had 55 blokes staying in hotel accommodation, at considerable cost to the union, together with a group of elite coaches who were concentrating their efforts on people of no obvious use to them. Wonderful.

My first thought as I drove back to town was something along the lines of 'this is nonsense, the whole thing is broken'. My second thought was that whoever might be to blame, it wasn't Andy. I knew how dedicated he was to his job: when you play against someone dozens of times at club level and operate alongside him for both England and the Lions, you learn a bit about him, and the thing that always struck me about Andy was the ferocity of his commitment to the cause of the moment. That being the case, I fully understood how frustrating this must have been for him. It wasn't as if he had too many aces to play in selection in terms of genuine world-class performers: he had a couple of leftovers from the 2003 squad and the odd gifted youngster, but there had been no system drive-through in terms of player development, for the very good reason that there wasn't much of a system. To be fair to Chris Spice, the Queenslander who had joined the RFU as performance director in 2001 and resigned a few weeks after the 2006 Six Nations, steps had been taken to improve the situation: an elite group of youngsters, including some of the best-known England players of recent times – Dylan Hartley, Danny Care, Danny Cipriani, Tom Youngs – had been identified and were being hot-housed when I arrived at Twickenham. But if the concrete had been laid, the construction work had yet to begin. These people were kids. The All Blacks were very definitely grown-ups.

The third thought that occurred to me was the bleakest, in human terms: Andy was on borrowed time. In fact, he was a dead man walking, terrible though that sounds. The same RFU hierarchy who had decided to keep him on after a disappointing 2006 Six Nations return had done for him after the subsequent summer series against the Wallabies, which had gone the way of the home side by the aggregate score of 77-21, with a try count of 9-2. Of course, there would have been a boardroom rethink had England gone through the autumn unbeaten, but as the New Zealand and Argentina Tests were to be followed by back-to-back meetings with the Springboks, the chances of a clean sweep were remote, to say the least.

It was not until the World Cup campaign of 2011 and its aftermath, which would reduce English rugby to a state of chaos, that the governing body rid itself of some of its more outdated practices, including the long-established old-school-tie, gentlemen's-club approach to hiring and firing national coaches. In 2006, some council members still treated Twickenham as their private fiefdom, even though the sport had been professional for more than a decade and there was meant to be some executive control over playing matters, just as there was over finance. Clive Woodward was given the coaching job in 1997 only after the union had failed to secure the New Zealander Graham Henry (who would go on to achieve a thing or two in the international game) and had at the same time lost the services of the incumbent, Jack Rowell, who understandably took a very dim view of being undermined by the pursuit of Henry and had decided to stick with his lucrative day job in big business. And when Clive's team underperformed at the 1999 World Cup, it was Fran Cotton, the chairman of the Club England committee, who did most to save his bacon.

Looking back, I was naïve in thinking that I would be part of a transparent, fully functioning chain of command, free of interference from county representatives with precious little knowledge of top-strata rugby. In those early weeks, it became clear to me that my job description and the reality of my job were not one and the same, and as the autumn series unfolded, I realised just how viperish a body the RFU could be.

The bright idea of playing the All Blacks out of schedule turned out not to be so bright after all: things started well enough when Tessa Jowell, the Secretary of State for Culture, Media and Sport, gave the new £100 million south stand its official opening by nailing a kick from in front of the sticks, but it went downhill from there and ended with a record 41-20 defeat at the hands of Daniel Carter, Richie McCaw and company. And then there was the fixture with the Pumas, one of those really gruesome situations in which Andy was at risk of being damned if he did and damned if he didn't. The South Americans had never won at Twickenham, but they won that day, partly because Toby Flood came off the bench for his debut and promptly chucked an interception pass in the direction of Federico Todeschini, but primarily because Argentina were the more cohesive unit. It was England's seventh successive defeat – a record-equalling slump rather than a record-breaking one, but grim all the same – and they were booed from the field.

If that was a painful moment, the after-match dinner was a

truly grisly affair that brings me out in a cold sweat whenever I think of it. There were calls from influential figures in the union for Andy to resign there and then. I was astonished. I said to Francis, 'Hang on just a moment. We're halfway through a four-match series, for Christ's sake. Let's just be sensible here. You have a head coach and a team who are ready to slit their own throats, they're that demoralised. And there are people here who want to make them feel worse?'

Unlike the vast majority of those who were running around demanding a change of coach within the next 20 minutes, I knew how the direct participants were suffering because I'd had first-hand experience of bad defeats in an England shirt. I'd lost 33-6 to Scotland in the days when tries were worth four points rather than five; I'd gone down 17-0 to Ireland in Dublin; I'd endured the misery of the inaugural World Cup in 1987. Where do they leave you, these humiliations? Emotionally speaking, they leave you face down in the ditch. Having failed to cope on the field, you can then look forward to feeling inadequate everywhere else: in the media room, over drinks with your family, at the banquet, and when reading the newspapers the following morning. Yes, the performance against the Pumas had been poor. Worse than poor. But I really didn't think this was the time to remove a coach's head from his shoulders. Not there and then. Not with the South Africans on our schedule for the next fortnight. Not with a World Cup less than a year distant and no succession planning in place.

This is not to say that I didn't recognise the gravity of the situation. The Argentina game was a perfectly accurate indicator, a true reflection, of where England were at as a Test team. Match days always are. You don't have to watch a team train to see what they amount to: it's all there in the 80 minutes, writ large. Rugby, cricket, football, whatever . . . If you know how to look, you find out everything you need to know about a side over the course of a contest. If there are problems of organisation, or confused thinking in selection, or issues on the practice pitch, or some kind of disruption through a breakdown in off-field discipline, it will reveal itself in performance. We were in a mess, patently, and it would take a lot of clearing up. I did not, however, believe that an overreaction to a particular moment in time was the way forward, even if the people doing the overreacting had been waiting breathlessly for an opportunity to wield the knife.

As coincidence would have it, there was plenty of blade brandishing going on in South Africa too, for the Boks had dropped a home Test to France, suffered the mother and father of a humping from Australia in Brisbane, and been convincingly beaten by Ireland in Dublin last time out. They were still dangerous – for all the torments of their Tri-Nations campaign that summer, they had salvaged something from the wreckage by beating both the All Blacks and the Wallabies at the back end of the series – and while they would be a much-changed side at the 2007 World Cup, they came to us armed with players as effective as Frans Steyn, Bryan

Habana, Butch James, John Smit, Danie Rossouw and Juan Smith. But the fact remained that their coach, Jake White, was under a good deal of heat, and when England recovered from a seven-point interval deficit to win the first match 23-21 on the back of a late front-rower's try from Phil Vickery, the temperature went clean off the gauge. Jake was called back to Cape Town to explain himself to the South African hierarchy, and when he arrived back in London to prepare for the second and final Test in the mini-series, he found himself one of the principal protagonists in a macabre 'him or me' drama. The feeling was that the coach finishing second at Twickenham on 25 November would be finished, full stop.

If the situation was not unprecedented, it was certainly uncommon, and it generated a tremendous amount of excitement. I can see that from the point of view of a mere onlooker, a disinterested observer, there was something thrillingly elemental about a contest between two men with their jobs on the line. It was certainly rugby in the raw. From my perspective, it was a very uncomfortable week indeed. We were desperate, they were desperate. Andy had his back to the wall, Jake had his back to the wall. And, in the end, it was Andy's wall that came tumbling down, thanks to four – yes, four – drop goals from André Pretorius, an outside-half from Johannesburg who had not started the previous week, but who made up for lost time by booting an international coach out of his job.

At ten o'clock that evening, the usual suspects were on the warpath. 'This is disgraceful and we're not having any more of it,' they were saying over their glasses of red. 'Robinson has to go. Get rid of him.' And that was it. End of story. I was left in no doubt that, before the weekend was out, there would be a vacancy at the top end of the England coaching operation. Having been through the post-Argentina frenzy and seen with my own eyes how these people were capable of reacting, I knew at the final whistle exactly what was about to happen. As did Andy, I think, but that did not make the following morning's phone call any easier. From memory, he was already back home in Bath when I dialled in with the bad news. It wasn't for him to ask me if the game was up. It was for me to tell him that the game *was* up. I didn't enjoy doing it, but the grandees had made it abundantly clear that my only choice was no choice at all.

Leaving to one side the most obviously unpleasant aspect of telling someone he is no longer in gainful employment, there were two things that irked me when I reflected on events. The first was a simple error: when I opened the morning newspaper and found that I'd been photographed sitting directly behind Andy at the final whistle, I cursed myself for not spotting the trap in advance and picking a different vantage point. It didn't look good, especially as there had been a fair amount of press comment to the effect that I was playing a role in England selection. This was a complete misrepresentation – I was the national head coach's immediate boss, not his puppet master; and in any case, no serious coach would accept being told who to pick – but the picture added fuel to the fire. The second problem was the lack of clarity over who controlled the hiring and firing. By going about things in the way they did, it seemed to me that a small group of RFU men wielded power without responsibility, creating the conditions for a change of coach and then saying: 'Right Rob, over to you. It's your job to sort it.' I should have made it clear, from the outset, that if all the tough stuff was going to land in my lap, I should be the one making the decisions. But I didn't, and I regret it.

Andy might just have earned himself a reprieve, or at least a stay of execution, had Pretorius not been so hot with his marksmanship and England had emerged from the autumn with a 2-0 victory over the Boks to set against the defeats of the opening fortnight. It might have been him at the 2007 World Cup rather than his South African opposite number. Would the Boks have won the title without Jake White? Would another Springbok coach have been able to persuade Eddie Jones to cross continents and add his rugby brainpower to the mix? We are in the realms of speculation.

What I knew at the time was that I had next to nothing to show for my first three months at Twickenham. Club and country matters? Barely an inch of progress. Age-group issues? Not even started. England? Worse than when I arrived. 'Things are going swimmingly,' I thought to myself. 'We're the defending world champions, the title goes on the line nine months from now and we can barely win a game. Our best players aren't delivering and there are no youngsters ready to challenge them in selection. The coaching team is up the spout, we have a dysfunctional relationship with the clubs and the governing body is all over the place. Right now, the good news is hard to find.'

It was at this point that I entered into discussions with Brian Ashton about succeeding Andy. As usual, the press was alive with ideas on who should be appointed. Nick Mallett, the English-born South African who had coached the Boks at the 1999 World Cup attracted some column inches, and there was a bandwagon of support for Shaun Edwards, the rugby league great who had worked wonders as a defence strategist at Wasps. But this was media-driven stuff, by and large; certainly, there were never any formal discussions with either man. Nick had publicly registered his discomfort at the idea of working alongside a rugby director. As for Shaun, it was widely accepted that he was reluctant to commit himself to anything more than a part-time role. For one reason or another, they could not be considered serious candidates.

Brian had a couple of advantages. He was already part of the back-room team, having been drafted in from Bath as attack specialist after the clear-out earlier in the year, and I knew he could coach. Everyone knew he could coach. I'd worked with him myself and admired his rugby intellect. Some of the most gifted players in the country swore by him, not least because he'd had a massive influence on them in a wholly positive sense. He'd also operated in the white heat of international rugby. Clive had drafted him in after the 'tour of hell' in 1998 and, over the next four years, he had sent England's attacking game into the stratosphere. They had gone through the 2000 Six Nations at four tries a game and upped it to almost six tries a game a season later. He had something, that was for sure, but was he the man to give us everything we needed?

It was hard to say for certain: I wasn't completely convinced that he would enjoy certain elements of a national head coach's role (my doubts in this area would turn out to be well founded), or that he had the really hard edge necessary to survive and thrive if things turned against him (and there were moments, particularly during the World Cup in France, when he was obviously stressed as hell). But, at the same time, the England set-up was in dire need of fresh thinking – the shock of the new, if you like.

We had arrived at one of those points where fundamental questions had to be asked about who we were and what we were trying to achieve. Brian was the kind of coach who could change the way players felt about their rugby, who could transform them in their hearts and minds, provided they had the imagination and the skill set to react to the challenges he set them. Under Andy, the environment had become claustrophobic: he came across as a growler, a snarler, a generator of huge intensity. Brian was the polar opposite. He was a master of freeing players up, of expanding their vision and injecting some enjoyment into training. As the autumn had been such a depressingly god-awful experience, maybe this was the way ahead: a swing away from what had just been tried; an antidote to Robbo.

And, crucially, he wanted to do the job. If I'd been completely upfront with Brian, I might have said: 'Okay, I'm giving you a hospital pass here. We can't be properly prepared for the very tricky Six Nations lurking just around the corner because the system doesn't allow it. Then we're off to South Africa for two Tests you can't possibly win, because all the best players are knackered already and you'll be travelling with a third team. And then there's the World Cup. Still fancy giving it a go? If you do, good luck. We're right behind you.' Instead, we talked it through in a more traditional fashion, and by the end of the conversation I was in no doubt that the desire was there. This was his moment, it might not come again, and he wanted to embrace it while he had the opportunity. Which was fair enough.

For the life of me, I couldn't see how we were going to make a fist of it at the forthcoming World Cup – the tournament by which we would be measured, every last one of us involved in the national set-up. How did we look, nine months out? I would have loved to have been able to say: 'We know our team, we're confident in our structures, we're a top-two nation with power to add, we're on an unbeaten run and we're flying.' I would also have liked to have described myself as a multibillionaire, but that wouldn't have been true either. The reality was as dark as could be. Very few educated rugby followers could be heard saying at the end of November 2006: 'You know what? You'll get to the final next year.' Come to think of it, absolutely *no* educated rugby followers were saying it. But at least we had a new coach in place – a coach who had familiarised himself with the demands of Test rugby, was full of innovative ideas, and had a happy knack of getting the best out of those players ready and willing to buy into a different approach to the game.

We would have made his life easier if, right from the start, we had allowed him to choose his own coaching team, but this was far less simple than it might have appeared. Following the departures of Phil Larder, Joe Lydon and Dave Alred in the post-Six Nations clear-out, the RFU had appointed three specialists to work with Andy Robinson: the former Leicester flanker John Wells, the ex-rugby league professional Mike Ford, and Brian himself. They ran the forwards, the defence and the attack respectively, and had been in post only a few months when Pretorius wreaked his havoc that day at Twickenham. No one pretended that Brian saw rugby the same way as John (and indeed their relationship would become very strained indeed as 2007 unfolded), but at the time of the regime change, there was little room in which to manoeuvre. We were already hard up against the clock. More change, even if it had been possible to negotiate, would have put us still further behind. This was not a moment for a further shuffling of the deckchairs. We were sinking faster than the *Titanic.*

It came as some relief, then, when we spent the first afternoon of the 2007 Six Nations on the crest of a wave. Jonny Wilkinson, finally hassle-free on the fitness front, returned to the side after a three-year absence for the meeting with Scotland at Twickenham, and promptly accumulated 27 points to eclipse by three the Calcutta Cup record I'd set in 1994: a terrific feat of 'full-house' scoring, even if he was so far in touch for his try in the corner that his feet were in Richmond High Street. Jason Robinson was also back, having decided that Brian's style of rugby was sufficiently invigorating to justify coming out of premature retirement, and there was a long-awaited debut in midfield for Andy Farrell, whose cross-code move to union had been blighted by injury. This was more like it: four crossings of the Scottish line, 42 points on the board, not even the faintest whiff of disgruntlement from the council members. That strange bright thing at the end of the tunnel – could it be a shaft of sunlight?

Of course it couldn't. It was a train, heading towards us at high speed with the name 'Croke Park' illuminated in the destination panel above the driver's window. If our second tournament match, an infuriating encounter with one of those Italian sides formidably equipped to win large amounts of possession but rather less capable when it came to capitalising on it, was something of a let-down, it was nowhere near the most alarming of the issues confronting us. The following weekend was fallow as far as the Six Nations was concerned, but instead of resting up ahead of the important game with Ireland in Dublin, the England players headed back to their clubs for a round of Premiership matches, some of which were scheduled for the Sunday. The Irish? They headed for their armchairs and a spell of rest and recuperation following their narrow defeat by the French. The playing field was about as level as the Wicklow Mountains and, thanks to the existing agreement between the RFU and the top-flight clubs, there was no way of flattening things out.

I was asked about the situation immediately after the Italy game and I made my feelings known to the media. 'If people play next Sunday,' I told reporters, 'we won't be able to train as a team on the Monday or Tuesday before the Ireland match. We'll be able to train on the Wednesday, but not on the Thursday because of the travelling. In effect, our single day's training will be set against the fortnight available to the Irish. It will be a massive disadvantage. All I can do is make a request to the clubs, but I'll be very surprised if there's a collective agreement. It's far more likely that decisions will be made on a player-by-player basis.' Which they were. Of the 22-man squad who faced Ireland on the last Saturday of February, no fewer than nine – Olly Morgan, our young full-back; Josh Lewsey, our World Cup-winning wing; Farrell, our brand-new inside centre; Joe Worsley and Magnus Lund, our starting flankers; and four of the replacements – had played league rugby the previous Sunday. Five others, including such important individuals as the scrum-half Harry Ellis and the lock Danny Grewcock, had been picked to play on the Saturday. There was a good deal of uproar, but while I was deep in negotiations with the clubs on a long-term deal designed to end this insanity, there was nothing to be done immediately.

Under the prevailing circumstances, that middle fixture of the tournament would have been awkward enough if we'd found ourselves playing relatively weak opposition in an atmosphere-free stadium. Against a pumped-up Ireland at Croke Park, of all places in the rugby universe, we were on a hiding to nothing. Opened in 1884, the stadium had been associated almost exclusively with Gaelic football and hurling, and for many decades there had been an official bar against the staging of sports seen as essentially foreign, primarily rugby union and football, which were termed 'garrison games' because of their connection with the British military. In 1920, during the Irish War of Independence, members of the Royal Irish Constabulary, supported by former British army officers operating as a counter-insurgency auxiliary unit, entered the ground during a Gaelic football match between Dublin and Tipperary and shot dead 13 spectators and a player in retaliation for a republican attack earlier that day. When, after the most delicate discussions, the modern-day Irish Rugby Football Union won permission to play a number of matches at 'Croker' while Lansdowne Road, their age-old home across town, was under redevelopment, the appearance of an England team on so politically sensitive a rectangle of turf was never going to be anything other than hugely symbolic.

The result has passed into history and is more than welcome to stay there. In the most emotionally charged stadium I have ever visited, we lost 43-13 – never had England conceded so many points in almost a century and a quarter of championship rugby – and were probably lucky to escape as lightly as we did. Afterwards, Brian said the England dressing room was 'like a mortuary', which was particularly apt, given that we'd crossed the Irish Sea with a death wish. I sat in the stand thinking: 'I'm part of the RFU, I'm one of the people responsible for this. Together with the clubs, we've just done everything within our power to undermine our own national team in a massive match in a huge competition in front of a vast television audience. We've basically said to them: "Now that we've made your lives as difficult as humanly possible, go out there and show us how good you are."'

Those 80 minutes in Dublin encapsulated everything that was wrong-headed and downright amateurish about English rugby. It reminded me, as though I needed reminding, that there was nothing clear-sighted or professional about what we had been doing since the World Cup win four years previously. The only positive I take from the experience now is that it concentrated minds, both inside and outside Twickenham. When my talks with the club negotiators resumed a few days later, there was a renewed sense of energy, of urgency. They knew, every bit as well as I did, that we were going to hell in a handcart and would bloody well burn unless we got our act together.

Recognition of a structural problem can happen in an instant: creating a structure fit for purpose is nobody's idea of an overnight job. Even though the discussions over the new club–country agreement were more focused in the aftermath of Croke Park, we were still some way off a deal. Meanwhile, there was the remainder of the Six Nations and the small matter of our two-Test jaunt to South Africa to consider. Brian reacted to the Irish humiliation by making 11 changes to his starting line-up, some of them forced by injury but by no means all, and the new-look team beat the French at Twickenham to end their hopes of a Grand Slam. There would be no follow-up victory in Wales in the final match of the competition, but with a couple of bright, young loose forwards on the flanks in James Haskell and Tom Rees, there were at least some signs of promise to come. However, the summer tour knocked us back again. Every bit as predictable as the shambles at 'Croker', it was every bit as painful as well.

With more than 30 players staying at home for one reason or another and another eight or nine confined to bed in the team hotel after contracting viral gastroenteritis – Simon Kemp, the England doctor, was heard to say that David Strettle of Harlequins, a recent addition on the wing, was 'as unwell as any fit 23-year-old rugby player I've ever seen' – the spectre of the damaging 1998 trip to the southern hemisphere was hanging over the group. But for some black humour, much of it coming from Brian's direction, it would have seemed like a two-month campaign rather than a two-week hop. On being introduced to the well-known sports broadcaster Jim Rosenthal, who would be presenting terrestrial coverage of the World Cup later in the year and had pitched up in Bloemfontein shortly before the first Test on a fact-finding mission, our new coach said: 'Bring your boots, did you? There's every chance you'll get a game.'

England lost both Tests, by an aggregate score of 111-32. The try count went 15-2 in the Springboks' favour. As the level of expectation was subterranean before departure, there was no enthusiasm at the RFU for another round of bloodletting on the squad's return. It was as if the only point of the trip was to fulfil our fixture obligations.

Yet few experiences in life are wholly devoid of value, difficult though the value may be to detect at the time, and we were able to take a modicum of encouragement, if not comfort, from this trip to the veldt. While only nine of the 29 players involved in the Bloemfontein and Pretoria Tests would be selected for the World Cup party (two more would be called in as injury replacements), the performances of a couple of players turned out to be highly significant. Brian was worried about the scrum-half position, where quality was at a premium – especially when the improving Ellis went down with a bad knee injury while playing for Leicester in their Premiership semi-final with Bristol and had to be wiped off the list of certainties for the global tournament. In South Africa, the much-travelled Andy Gomarsall was restored to national colours and played well enough to bag himself a World Cup place. Something similar happened with Mark Regan, the hale and hearty hooker from Bristol who, thoroughly hacked off at being jettisoned from the England side by Andy Robinson, had retired from Test rugby in 2004. 'Ronnie' Regan respected Brian; he readily agreed to fill the hole that had opened up in the middle of the England front row and played himself to a standstill. He too would be a key figure at the World Cup.

That tournament was played out over six weeks or so and, in its own way, matched the 1995 competition in straining credulity to the limit. My role here was less active in the physical sense, but every bit as exhausting in the mental and emotional ones, which was partly because the negotiations with the clubs were at a sensitive stage and I was going backwards and forwards from France in an effort to manage two ongoing situations of enormous significance to the game in England, and partly because the changing fortunes of the team gave everyone associated with it something akin to an out-of-body experience. For a side to perform like a pub team in the opening pool game, against the United States in Lens, and then perform an uncannily accurate impersonation of – let's be blunt – a pissed pub team against the Springboks in Paris a few days later, yet still find their way to the final: it was not, to say the very least, an everyday tale of rugby recovery.

We were all but dead and buried after losing 36-0 to the South Africans, a classic instance of the losers being lucky to get nil, yet we rose again to make it into, and through, the knock-out phase. Sitting in my posh seat in the Stade de France for the final, a winner-take-all showpiece rematch with the Boks, I could not help wondering if Phil Vickery, our captain, had taken his tactical briefing from Brian, or from Lazarus.

If the 36-0 night was desperate, the immediate aftermath was little better. The press, never slow to kick a head coach when he's down, were really getting stuck into Brian, and when news leaked out of a behind-closed-doors, heart-to-heart session involving the coaches and players, there was a small avalanche of stories about senior members of the squad taking control of the situation, leaving the man in charge with a title but no role – stories that grew in both number and momentum as England began, in the face of mighty odds, to make themselves relevant again.

I did not attend the meeting in question, which was held the day after the humiliation at the hands of the Boks. I'd flown back to England that Sunday for another round of discussions with Tom Walkinshaw and the other club negotiators, and didn't return to France until two or three days before the next game, a must-win affair with the dangerous Samoans. As a consequence, I was left with only a second-hand impression of what happened over the course of that so-called 'clear-the-air summit' and never felt I got to the bottom of it. Different players saw the events in different

Brian Ashton and Jonny Wilkinson chew the fat ahead of the 2007 World Cup, which we could have won against astronomical odds. Lawrence Dallaglio, Mike Catt, Mathew Tait, Jonny, Andy Gomarsall and Andrew Sheridan ponder what might have been after losing a tryless final to South Africa.

The 2011 World Cup had trouble written all over it. Mike Tindall, above, and Manu Tuilagi, above right, were two of the principal protagonists.

I join Martin Johnson in the firing line as the press take aim at England's hierarchy.

New coach Stuart Lancaster, in the white top, with his three wise men: Mike Catt, Graham Rowntree and Andy Farrell.

Brad Barritt slams the door on the predatory Julian Savea.

Wales in Cardiff? Big problem. The Grand Slam game in 2013 ended in a defeat of record proportions and was a chastening moment for anyone with an English accent.

Deflated and downcast, England were no match for the Wallabies a week later.

ictory over the less than mighty Uruguay in a dead fixture seemed meaningless at the time, though it marked the start of a record-equalling run of victories. It also marked the end of tuart Lancaster's coaching tenure and Chris Robshaw's spell as captain.

The smile of confidence. Eddie Jones, an obvious choice as Lancaster's successor, has never been prone to self-doubt.

The new coach surveys his Twickenham fiefdom with RFU chairman Bill Beaumont, left, and chief executive Ian Ritchie before presiding over an immediate victory in Scotland.

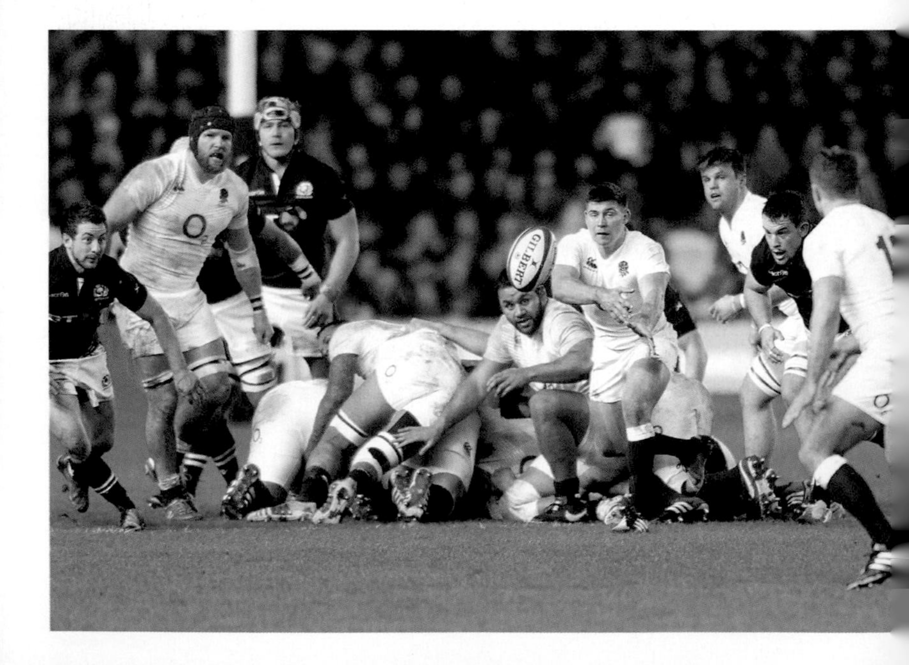

ne Six Nations, one Grand Slam. England did the necessary against France in Paris to complete
clean sweep and marked the moment in time-honoured style.

ere was no stopping England in 2016. Dylan Hartley led the team to a ground-breaking
ies victory in Australia and followed up with a comprehensive win over the Springboks
Twickenham.

The English production line has been pumping out Test-quality youngsters since the agreement between the RFU and the Premiership clubs in 2007. Maro Itoje and Owen Farrell all made major contributions as the British and Irish Lions fought the All Blacks to a standstill in 2017.

My new role at Sussex CCC requires a business suit and a freshly ironed shirt rather than glove and pads, but when it comes to watching the cricket at Hove, I have the best seat in the house.

ways and drew different conclusions, inevitably driven by their standings in the squad pecking order, their views on tactics and selection, and their personal agendas.

It was, and still is, easy to see both sides of the argument that dominated so much of the media coverage between our two encounters with the South Africans. I'd been in plenty of meetings as a player, some with coaches present and others with coaches absent, where we'd decided to ignore the tracksuit brigade and do it our way. After England won the World Cup under Clive, there were plenty of tales about certain players slamming the door shut and saying: 'Right, sit down and listen. Forget everything you've just heard from the boss. We're doing this, this and this. We're not doing that or that. Clear?' During their trophy-laden days at the Recreation Ground, players at Bath could be similarly dismissive of Jack Rowell, despite his 'great man' image in the minds of the rugby public. 'Jack's in one of his strange moods today,' they would say. 'Let's agree not to listen.'

There might well have been an element of player power, for want of a better description, in that post-Springbok heart-to-heart, but there again, Brian was precisely the kind of coach who believed in player empowerment – who went out of his way to encourage a sense of independence and responsibility among his charges. 'What did you expect us to do after a 36-0 beating at a World Cup?' he asked when pressed on whether he was still calling the shots. '*Not* talk about it?' There were things that needed sorting – that much was blindingly obvious to anyone who had clapped eyes on England at any point during the first two matches – and Brian was not the kind of character who could easily persuade his critics that things were not as bad as they seemed or conceal his stresses and strains beneath a veneer of authority. But in truth, no coach in my experience was ever in complete control of a team: not Geoff Cooke, not Jack, not anyone I'd ever played for at club level. And in major tournaments and Test series, that tends to be doubly true. Half the players feel marginalised because they're not in the team and it doesn't take much for them to air their grievances, especially when they're drowning their sorrows in the local bars – something that always happens, without fail.

Samoa week was an interesting one, that's for sure. The talks with the clubs were advancing, but we were still a little way off an agreement: in fact, after the 36-0 hiding on the other side of the Channel, whatever feel-good atmosphere had been generated around the table seemed to evaporate. When I rejoined the England camp for the big game in Nantes, they were preparing at La Baule-Escoublac, about an hour's drive to the west of the city on the Brittany coast. I spoke to Brian, I spoke to the senior players; I tried my best to get a handle on what was happening. I did not emerge from those discussions feeling particularly good about life. The overwhelming feeling was one of pressure – hot, sticky, claustrophobic, energy-sapping, morale-deadening, paralysing – and I knew the best organised of the Pacific islands teams would be highly motivated. (They had already lost twice but there was still a quarter-final route open to them.) Defeat would almost certainly eliminate us from the tournament with a game to spare. Terrific.

There were times when I travelled to an international match on the players' bus, but on this occasion, I was invited to share a car with Francis Baron. It was not an invitation I felt I could refuse, so off we headed on our 50-mile drive inland. Every time Francis asked me what I thought was about to happen, I came up with the same profoundly unenlightening answer. I did not have even half an idea. Jonny Wilkinson, injured before the opening game when his attempt to evade the 18-stone lock, Steve Borthwick, during what was meant to be a non-contact training session had left him nursing a damaged ankle, was back in the side, which was a welcome bonus. But England had no form behind them and the nature of their predicament was not lost on the more vulture-ish members of Her Majesty's Press, who, smelling carrion from afar, had swept across from Fleet Street in threatening formation and perched themselves on the media benches in expectation of a feeding frenzy. The one thing I felt I was able to tell Francis for certain was that, if England lost, no one would escape the backlash in the newspapers. Not me, not him, not any of us.

Looking back on it from a decade's distance – which is quite close enough, thank you – I remember the game as a real stomach-churner. Before kick-off, I attended a pre-match function where Bill Beaumont was speaking to a high-powered audience of union administrators and financiers about the glorious unpredictability of World Cup rugby and the dramatic potential of the upcoming fixtures, ours included. My response was not a generous one. 'Never mind all this crap,' I recall thinking. 'I just want this bloody thing won.' And win it we did, although there were moments in the second half when the Samoans were on a surge and I was transfixed for all the wrong reasons, like a man of faint heart watching a horror movie through his fingers. It was then that I recognised the truth of the old saying about rugby being easier to play than to watch. As a coach, you're at one remove on game day – wholly responsible, yet essentially powerless. As an administrator, you're doubly helpless. When the Irish referee Alan Lewis blew for time, the sense of relief was overwhelming. Somehow, we were still alive.

The story of the tournament from there on in has been told on thousands of occasions, in almost as many different ways. But facts are facts: we upset the odds in the last eight by beating Australia in the way England sides generally beat Australia – that is to say, through hard, relentless forward play, mixing aggression and control in equal parts. (One of the upsides of that particular performance was the positive effect on all of us around the club–country negotiating table. We reached agreement within days of the victory.) We then saw off France in the semi-final a little more comfortably than the five-point winning margin suggested, before losing a tight, try-less final to the Springboks, having matched them pretty much everywhere but in the line-out, where we were comprehensively destroyed. With a little more possession, we might have become the first nation to successfully defend a world title. Had we done so, I would have been tempted to wonder whether Brian had manufactured that 36-0 defeat as part of his tournament strategy.

The days and weeks after the final were strange indeed. As director of elite rugby, I was immediately pressed on Brian's future by those in the media who, it soon became clear, were determined to see the back of him, despite the fact that we'd stayed in the competition longer than the All Blacks, the Wallabies, the French and all three Celtic nations. There was never a serious possibility of him being pushed out immediately, but as there was a formal review process to be conducted, I wasn't in a position to say so categorically.

The problems intensified after we reconfirmed him as head coach. Brian wanted to make changes to the coaching panel, and we should have allowed him to do so. I know that now. It was a perfectly reasonable request from a man convinced of his ability to raise the team several levels if he was given the freedom to work with like-minded colleagues who were truly comfortable with his rugby philosophy. So why did we stick with those already in place? I don't mean this as an excuse, but one of the disadvantages for the northern hemisphere sides in this regard is the limited time between the end of a World Cup and the start of a Six Nations.

Win or lose, World Cups are watershed events that invariably expose areas of weakness in a national set-up, leaving an awful lot of things up in the air from an administrator's point of view. Just when you're thinking, 'Thank Christ that's over', you hear a small voice saying: 'Yes, but it's Wales in a couple of weeks. God help you if you lose.' The New Zealanders, the Australians and the South Africans can put their feet up until after Christmas, happy in the knowledge that they won't be playing their next Test until June. Up here in Europe, the must-win matches just keep on coming. So while I could see Brian's point about new faces on the staff, there also seemed to be a powerful argument in favour of continuity.

Would things have unravelled for him over the course of the fateful 2008 Six Nations if he had been able to choose his lieutenants? Possibly not. There again, he could have had the late Carwyn James and Merlin the magician by his side and still been powerless in the face of events that led to the Welsh victory at Twickenham in the opening match. England were comfortable enough until the injuries kicked in: so many injuries, indeed, that a cavalcade of ambulances left Twickenham at close of play, all of them bearing stricken men in white shirts. Deprived of a functioning back-row combination for the entire second half, we conceded late tries to Lee Byrne and Mike Phillips and came up seven points short. When, eight days later in Rome, we almost went down to Italy, I had the uncomfortable feeling that Brian would find himself in Andy Robinson-land unless he could work the oracle pretty damned quickly. Which he did to a degree, conjuring another win over the French in Paris – a match in which Jonny drew level with Neil Jenkins of Wales as international rugby's all-time leading scorer. But it was a short-lived reprieve: in the middle weeks of the competition it became obvious that some very powerful figures had decided that whatever journey Brian was on, it was not one they intended to share.

I might have been at the RFU a mere 18 months, but it was comfortably long enough to recognise the sounds and smells of a Twickenham plot. My senses were in overdrive when, in the penultimate championship match, we lost to Scotland at Murrayfield by five penalties to three. (A veritable feast of running rugby, as dear old Bill McLaren would certainly not have said.) It really kicked off in the post-match function room, with a number of disgruntled council members engaged in rounding up support for a move on Brian. There were a few committee men I trusted to keep an ear to the ground and fill me in – they confirmed that there was the scent of a kill in the air. Yet no one came to me during the week that followed with a 'get rid of him' message. Instead, the first conversation I had on the subject was with David Hands, the long-serving rugby correspondent of *The Times*, who, on the day before the last match with Ireland, phoned me out of the blue and said: 'I understand Brian's days are numbered.' I stonewalled. 'Well, I haven't had any discussions on this,' I said truthfully, albeit in the narrowest of senses. 'The Six Nations hasn't finished yet. There'll be a review when it's over, because there always is.' My heart sank a little. I hadn't enjoyed the Robinson episode one little bit, and the signs were that this would be worse.

Against Ireland, Brian's bold decision to drop Jonny to the bench and give a young Danny Cipriani a chance to bring some pizzazz to the mix paid rich dividends in an excellent 33-10 victory, but the fact that we'd played extremely well did not amount to a row of beans in the great scheme of things. I was not ordered by Marytn Thomas or Francis, in so many words, to sack Brian: in the alternative universe of the Twickenham hierarchy, things were rarely quite so clear-cut. Rather, word was being spread that Martin Johnson, less than three years into retirement and no one's idea of a career coach, might be interested in succeeding Brian if asked, and that the RFU was more than interested in doing the asking. Where did all this come from? It was never entirely clear to me that it was a union production pure and simple. There seemed to be a big Leicester drive behind the pro-Johnson campaign and I wondered at the time whether Peter Wheeler might have been one of the moving spirits. He, like Martin, was a Tigers man through and through, and he held his fellow club legend in the highest esteem. Whatever the truth of it, I was genuinely taken by surprise, both by the speed of Brian's brutal rejection by the hierarchy – I can put it no differently, for that is what it was – and by the identity of the favoured successor.

Of course, I knew there were things in need of addressing. If the Wales defeat had been a classic example of sod's law at work, Italy had been grim and Scotland had been dire. But post-World Cup periods always presented a challenge; Brian had been denied a coaching team cast in his own image; and if everyone who lost a game at Murrayfield was immediately declared persona non grata, the English rugby landscape would be thinly populated indeed. My chief sadness over the Ashton affair is the desperately poor way in which it was conducted. It was a real mess. No one from the RFU came even close to handling it well, and I include myself in that. There was still a grey area over how decisions on the England coaching staff were made and who was responsible for making them, but whichever way you cut it, Brian was hung out to dry. I have to take my share of the blame for the way it happened.

So there I found myself, talking through the job with Martin Johnson, while not being wholly honest or straight with Brian, who was still in post and mulling over changes he wanted to make ahead of that summer's two-Test trip to New Zealand. I was uncomfortable with the prospect of appointing Johnno, not because I had any issue with him as a bloke, but because Brian was being left dangling, and for a while I assumed that Martin would, after due reflection, decide that in the complete absence of coaching or management credentials, it would be better if he didn't get involved. Indeed, over the course of several meetings with him and Tim Buttimore, his agent, there were moments when it seemed to me that Tim himself had reservations. But while it was difficult to know exactly what was going on behind that famously impenetrable Johnson façade, it slowly became clear that he felt he had something to offer. And that, in the end, was all Martyn Thomas and his fellow Twickenham grandees wanted to hear. I was part of a delivery process, and I duly delivered Martin to the RFU, as requested. But I rarely go more than a few weeks without thinking back on that affair and regretting our behaviour as a union all over again.

9
THE TOUGHEST WEEK

AT SOME POINT during the afternoon or early evening of Sunday 15 June 2008 – I forget exactly when, probably because I've locked away the precise detail in a compartment of my mind reserved for those episodes in my rugby life I have no great wish to remember – I had a conversation with Richard Smith QC, a barrister from the West Country who had, since Clive Woodward's time in charge of the national team, been England rugby's travelling lawyer: our go-to man for all matters regulatory and disciplinary. We were halfway through a two-Test visit to New Zealand and had just flown from Auckland, where we had lost 37-20 to the All Blacks at Eden Park the previous evening, to Christchurch, where the second match would be played at Lancaster Park a few days hence. It had always looked like being a tough tour; it was turning out to be a tough tour; and by God, it was about to become a whole lot tougher than anyone could conceivably have imagined.

Richard had picked up some news from one of his legal contacts in the country that was alarming to say the least:

various journalistic outlets had become aware of an allegation of sexual impropriety made against a number of England players, and there would be a significant amount of coverage in the Monday morning papers and bulletins. The difficulties of the Six Nations, the painful ravages of the Brian Ashton affair and an absentee list of injured players as important to us as Danny Cipriani, Jonny Wilkinson, Phil Vickery, Simon Shaw and Nick Easter – not to mention the high quality of the opposition – had put us in damage-limitation mode from the outset. Little did we know that the eventual extent of the damage, and the challenge of limiting it, would be so great. By the end of the trip, the problems posed by Richie McCaw, Dan Carter, Conrad Smith and the rest of the silver-ferned aristocracy seemed wholly irrelevant. We had other issues on our minds.

I was leading the tour. The honour, if that is what it was, should have been Martin Johnson's, but during my clandestine discussions with him over the course of the Six Nations at a variety of venues between London and the Midlands (often at a hotel near Market Harborough, just off the M1), it became clear that he would not be in a position to take on any active England role until midsummer. His wife Kay was expecting and that was his priority. Fair enough. It was only as events unfolded in Christchurch that I found myself accusing him, wholly unfairly, of throwing the biggest sidestep in the history of second-row forwards.

Did I want to fill in as caretaker? Not really. Was it

appropriate that I should have done? There are arguments either way, but it was certainly true that having been caught up in the Ashton controversy in so public a fashion, the situation was some way short of ideal. Such are the complications of life in high-end sports administration. Had things turned out differently at Murrayfield a few weeks previously, Brian would probably have stayed in place for a while longer – long enough, at least, to take the side into All Black country. And what if Martin had turned down Twickenham's approaches and opted to stick with the quiet life? That possibility lay at the heart of what came to be seen as our shabby treatment of Brian. It is not easy to come clean with an incumbent while the man you're courting as his replacement is considering his position and may end up knocking you back. We all know of circumstances in football where a manager is sacked on a Tuesday afternoon, his successor is unveiled on the Wednesday morning and everyone says: 'Crikey, that was quick. Anyone would think it had all been set up in advance.' And to add to the delicacy of this particular set of discussions, there was a chance that Brian would be asked to stay on in a pure coaching capacity, under Martin's management. Martin didn't seem in favour of that idea as it turned out, but there was a moment when it seemed possible.

With Martin signed up for the next England journey only in the metaphorical sense (he would select the squad, then watch the action from afar), there was a choice to be made. John Wells and Mike Ford were still in place as specialist coaches, as was Graham Rowntree, who had come on board as a scrum expert during the 2007 World Cup and had been of considerable value since. Jon Callard, alongside whom I'd played towards the back end of my international career, also had a role as kicking coach. Was it reasonable to ask one of the senior back-room staff, John or Mike, to front things up in Martin's absence: to take on a fresh set of responsibilities, including the media chores, while attempting to draw the sting from a motivated New Zealand side still smarting from another failed attempt to reclaim the global title? All things considered, it would have been an unfair demand. I probably should have done more to avoid undertaking the task myself, perhaps by playing the Johnno card and claiming paternity leave, but there were no hot ideas on alternatives from the grandees at the Rugby Football Union, who, having quickly put the unsavoury Ashton business behind them, were looking to the professional staff to deliver an immediate upturn in fortunes. It was ever thus.

When I reflect on the trip now, the best part of a decade on, I still find parts of the experience difficult to comprehend. I find it equally hard to believe that only three years later, at a World Cup in the same country, another set of England players found ways to present the scandal-mongering wing of the journalistic profession with gifts from the gods. There were many points of difference between the two situations, not least the gravity of the accusation against the 2008 group, but there were similarities in the sense that a bunch of professional sportsmen on a 'big night out' were found badly wanting in the crucial field of self-discipline. You can lay down as many laws as you like about tour standards, and employ every security guard on the planet in an effort to save players from themselves, but unless those players have a sense of their own vulnerability, unless they understand the importance of self-control in the face of a thousand temptations, there is no guarantee of safe passage through a rugby weekend. I'm not a complete puritan in these matters: the England teams in which I played were more than capable of letting rip themselves. But 2008 had precious little in common with 1988 in terms of rugby's worldwide profile, and as social media continues to evolve, the long-established practice of sweeping controversy under the carpet will be nothing more than a distant memory by the time we reach 2028.

The Eden Park game had been played during the evening, as is usual in New Zealand these days, and once the after-match formalities were done and dusted, a number of players left our harbourside hotel and made the short trip into town. There was nothing unusual about it: they knew there would be only a light recovery session the following day, much of which would be eaten up by the journey to the South Island, and anyway, those of us on the management team had taken steps to ensure they would be enjoying a drink or two in a relatively controlled environment. Yes, this was the most rugby-obsessed country on the face of the earth and, therefore, the last place on earth where an international rugby player might consume a few beers without being spotted; yes, there had been a history of tour parties being targeted by the local press (just ask the 1977 Lions about the blazing front-page headlines accusing them of being 'lousy lovers'). But would it really have been a good idea to slap a curfew on the entire squad and risk making them feel as though they were atoning for past sins? Would it have been right to deny them some convivial downtime? I didn't think so. During England weeks in the recent past, Brian Ashton had routinely sanctioned an hour or two of bar activity for those players who wanted to indulge. 'Just don't abuse the privilege,' he would say. It seemed a perfectly sensible, grown-up way of going about things.

In Auckland, unfortunately, some of the younger players allowed a straightforward night out to mutate into a daft night out: instead of staying in Bar A or Bar B, as planned and agreed, they hit Bar C and Bar D and on through the alphabet. And when they eventually returned to the hotel, they had female company. We had put transport and security in place, but it was as nothing compared to the modern-day arrangements, where the people employed to manage things on the ground are basically told: 'The players will be back in this hotel by such and such a time. If they don't want to come back, it's your job to persuade them. And if they won't be persuaded, you pick them up, stick them over your shoulders and don't put them down until they're safely in the lobby. Yes, that includes the tight-head props.' Back in 2008, the system was comparatively primitive. What was more, it was an unfamiliar squad, with little in the way of a team culture to underpin it. There could not have been a captain more aware of the importance of discipline than the ultra-serious Steve Borthwick, but he had only just been appointed to the position; it was hard to think of a new boss who commanded more respect from the English rugby public than Martin Johnson, but he was back home in England. For this brief period, everything was up in the air.

Having been forewarned by Richard Smith, I woke up on the Monday morning in a state of nervous anticipation. Sure enough, the news had broken. Initially, there was significant confusion over what exactly was being said against our players, but the story moved on quickly in the local media – plainly, there was a good deal of information emerging from police headquarters in Auckland – and it was not long before we found ourselves dealing with an allegation of rape. We were given the names of four players privately identified by the police as people they were keen to question, so we spoke to each of them before calling a meeting of the entire squad for nine a.m. We explained to them what we knew: we said there was a really serious allegation being levelled at some of the people in the room, but that we didn't have all of the facts and would have to find out as much about what had happened as quickly as we could. We warned there was going to be an absolute shit-storm in the media and told the players that none of them should say anything to anyone on the grounds that if they did, there was no knowing where the information would go and what effect it might have on those involved. We had to lock it down.

Quite how any of us would have survived the week without Richard, I have no idea. He was utterly brilliant. When I was wearing the England No. 10 shirt, the only way I'd have found myself travelling with a QC – or a lawyer of any description, come to that – was if he'd been a good enough player to be picked for the squad. A decade or so on, a legal advisor had become something more than one of rugby's trendy accessories, however much the southern hemisphere countries mocked England for including a 'man with a wig' in their tour parties: he was now considered an essential part of the set-up. The regulatory framework within which international rugby was played in the amateur era had its share of complexities. In the professional age it had quickly become a minefield, especially at major tournaments where the difference between a one-match ban and a three-match ban slapped on a single player could also be the difference between success and failure – between a coach remaining in his job or being handed his cards. During the 2003 World Cup, when England somehow contrived to have 16 players on the field for half a minute or so during the fraught pool game with Samoa, it was Richard who quickly saw the threat of a possible points deduction and found a way to head it off at the pass. He had more than justified his travel expenses on subsequent tours, too. But this situation was different. This was not about rugby. This was serious.

Each day, the pressure on us intensified as new information appeared in the media, fresh challenges were created and the dramatis personae expanded to fill the room available. We were in close touch with lawyers both in Auckland and Christchurch and as a result of those discussions, we were adamant that the flow of information from our side be kept to a minimum, because none of us could predict the direction in which this sorry business would move. We were aware that the police could not insist on interviewing members of the tour party in the absence of a formal complaint from the alleged victim, and as no such complaint had been made, there was no reason for us to make any of our players available to officers on a voluntary basis. But that was only half the story. The English newspapers, never slow to pick up the scent of a full-bore sporting scandal, were sending their hard news types – Fleet Street's Rottweiler breed – all the way to New Zealand in search of a 'proper yarn'. Meanwhile, members of the RFU council back home were demanding action, even though no one knew exactly what kind of action they wanted to see; and Francis Baron, the governing body's chief executive, was arguing that we should allow him to say something statesmanlike. I had people from Twickenham on the phone, calling for an immediate RFU inquiry. In response, I was telling them to hold on and allow those of us on the ground to handle the situation – a situation so fluid and dynamic that I was meeting Richard every hour on the hour from early morning until late in the evening.

We were in uncharted territory and by way of a sideshow, we had the small matter of the second Test with the All Blacks looming before us. I said to the coaches: 'Look, I know this is impossible, but we need to get a team on the pitch on Saturday come what may. Just do what you can, chaps. Thanks.'

We took the view that we should stick rigidly to legal process and if that left us open to public criticism, so be it; and indeed, there was plenty of comment to the effect that members of the England management team were being outrageously obstructive in refusing to help the police with their inquiries. In our view, there was no logic in doing anything else: there were more lawyers swarming around than you're likely to see on a busy day at the Old Bailey, and the collective advice was to sit tight and wait for the detectives to produce the required paperwork. Which they never did. It came to a head when some officers flew down to Christchurch from Auckland in order to interview the players. They knew they didn't have the authority and they knew that we knew they didn't, but they came anyway. When they arrived, we refused them access on the guidance we'd received. It was a pretty ballsy thing to do but we had questioned the players, in great detail, and believed their accounts of the incident.

Richard was the most reassuring of figures. Had he not been there to lend his expertise, the words 'up', 'creek' and 'paddle' would have been the first to pass my lips, along with 'help' and 'panic'. I would not have known where to start, where to turn, what to do, who to consult. But it was not until we finally left the ground at Christchurch Airport and the nose of the plane turned towards Heathrow that I felt I could breathe properly. Right the way through that horrible week, I was acutely aware that if the complainant suddenly gave the police what they needed, there would be some very loud knocking on our doors. Imagine going to bed every night in the knowledge that, by morning, a bad situation could be immeasurably worse. Imagine being one of the players. They were petrified. Yes, there had been some stupid, self-destructive behaviour, yet they continued to train and, in some cases, faced the All Blacks for a second time. They might have put themselves in a world of trouble off the field by going to the wrong kind of place at the wrong kind of time, but I could not question their commitment as rugby players.

As match day drew closer, the media net drew tighter. The recognised rugby press corps were staying in the team hotel, which just happened to have been constructed on an open-plan basis with a nice big lobby and acres of public space: brilliant for the journalists, not so clever for an England team trying to keep as low a profile as possible. Even the reporters we knew were under instructions from their editors back in London to unearth a tasty detail or two, badger their contacts for fresh information, and above all, to identify the players involved. The reporters we didn't know were another problem entirely. Some of them were back in Auckland, talking to hotel workers and bar staff and flashing the cash in pursuit of something lurid. If there was pressure on us, I hate to think how grim that week must have been for the girl and her family, with what seemed to be an entire country waiting for . . . who knew what?

On the eve of the Test match, Francis had his moment in front of the media. I felt he struck the right note, successfully balancing the frustration of the tour party at the unsatisfactory nature of the legal limbo – 'These young lads don't know how to get to a situation where their names can be cleared; in the absence of a complaint there appears to be no immediate end to this process,' he said – against our own sense of regret that we should have been the source of so much trouble and embarrassment. 'Over the years, players representing England teams have been outstanding ambassadors for the game and for our country,' he continued. 'We've never had an incident like this and we never want to have one again. These things are very damaging. Clearly, we have to get across to these young men that they are targets. They have to be so conscious of what they do and how they do it. We have to make sure . . . that we educate our young men in the realities of the new world of celebrity.' And he also went on the offensive, railing against the media for buttonholing the families of various members of the squad. 'We are hearing reports of mothers being approached,' he said, the anger rising in his voice. 'That's why it is so important to bring closure to this incident. If a complaint is not made in the next two or three days, I'm sure our legal team will be pressing police to close the file. The damage to the individuals is very substantial and that is unfair under any jurisdiction. There are some lurid allegations floating around.'

Those allegations would continue to float for some time, but when Saturday dawned, at last there was some blissful distraction in the shape of a game of rugby. In the great scheme of things, the match was neither here nor there: the problems of the last week were unlikely to be alleviated, let alone solved, by the fact that 30 blokes were about to spend 80 minutes knocking lumps out of each other. But it provided a timely reminder of what normality looked like: not normality in the sense of men in black shirts registering a comfortable victory over men in white shirts (although that is certainly what happened), but normality in the sense that the game had not changed out of all recognition in the space of a few days. There were scrums and line-outs and thumping tackles and tries, two of them scored by us. We lost 44-12 but, somehow, the scale of the defeat did not hurt as much as it might otherwise have done.

Back we went to our hotel-turned-prison, with its lovely big lobby and open-view bar providing all and sundry with a perfect vantage point from which to observe our behaviour. To the surprise of some people – maybe most people – a few of the players were keen to head into town for a drink or two, and we in the management were open to the idea. Could this possibly have been a wise move, given what we had just been through as a group? There had been a serious discussion about it, of course, but it had been a rough tour on pretty much every level; everyone had taken just about as much as they could take and each individual was in need of some kind of release. The players directly implicated in the police investigation had been through hell, but events had taken their toll on all of us: many of those players not involved had spent the week in a state of considerable distress – understandably, for there were some strong friendships within the group. Some members of the squad decided to stay in the hotel, even though there was little enough entertainment on offer; others popped up to their rooms, put on a smart shirt and went in search of a pub. There was no major address on the 'dos and don'ts' from anyone in the management or from the RFU. Just a mild, gently ironic reminder along the lines of: 'Whatever you do, please try to behave. There's a flight home tomorrow morning and it would be nice if we were all on it.'

Me? I spent an hour or so alone in the hotel bar, nursing a beer and chewing over the problems that still lay ahead. For a start, there was still a possibility of some dramatic police intervention: our players could yet feel the dreaded tap on the shoulder while sitting in the departure lounge at the airport, or even standing on the steps of the plane. There would also be a full RFU inquiry into the events of the tour, an investigation that was certain to prove extremely uncomfortable for everyone involved.

That inquiry was conducted by Judge Jeff Blackett, the union's disciplinary officer, and he didn't hang around. We touched down at Heathrow in the last week of June; his report, very full and very frank, was in the public domain midway through the second week in July. Two players, the Harlequins full-back Mike Brown and the London Irish wing Topsy Ojo, were found guilty of misconduct – Brown for staying out all night and arriving late for a physio appointment; Ojo for simply staying out all night – and fined £1,000 and £500 respectively. Two other players were cleared of misconduct. Jeff was sharp in his condemnation of what he considered errors of judgement by a number of squad members, but he also rounded on parts of the media for 'irresponsible and wildly inaccurate reporting'. Crucially from the point of view of those players who had been of most interest to the Auckland police, he stated: 'All the players I have interviewed vehemently deny any criminal wrongdoing and I have seen or heard no evidence which has been tested to gainsay those denials.' Finally, it was over. If I hadn't been so tired, I'd have jumped for joy. It had been a thoroughly miserable business and, while it had lasted less than a month from start to finish, it had dominated every waking hour of my life during that time. Can 26 days really seem like a lifetime? You bet it can.

If the sense of relief lasted me through the summer break, there was more than enough work piling up on my desk at Twickenham to generate new pressure through the autumn. I was no longer the England team manager, thank heaven. At a slightly uncomfortable final press conference in Christchurch before the ride to the airport, Eddie Butler of *The Observer* had asked me if I was planning to resign. Resign from what? From a job that hadn't been mine before the tour and wouldn't be mine by the time we returned to England? I was certainly not planning to walk away from my proper job, which still carried the title of Elite Rugby Director, and would do so for a while yet, even though I was picking up faint warning signals of hassle to come. There had actually been some rugby played on our trip to New Zealand, not that anyone had taken much notice of it once the extra-curricular activities had hit the headlines, and it required a proper review. Following on from that, there was Test business at Twickenham on the horizon: matches against the Pacific Islands – a select team featuring the best players from Fiji, Samoa and Tonga – and the three traditional big-hitters from the southern hemisphere. Another meeting with the All Blacks? Lovely. Who could ask for more?

There was a good deal of talk about the shape of our summer tours and the value we were extracting from them and, on the face of it, people were justified in their concerns. Since the professionalisation of the sport in 1995, we had faced the Australians, the New Zealanders and the South Africans in tour matches on their own soil on 19 occasions, lost 16 of those matches (most of them by a distance) and conceded more than 700 points while scoring fewer than 250 of our own. It was only when we headed west rather than east in British and Irish Lions years and took on Argentina or the North American nations that we travelled in expectation as well as hope. Yes, we often went with understrength squads; yes, our players were pretty much on their knees at the end of long, bitterly fought club campaigns. But as our critics never tired of pointing out, there was no hint of the pushover about the Wallabies, the All Blacks or the Springboks when they flew north in November.

If those Test matches were generally more competitive, England still found themselves on the losing side more often than not. 'They have a long season too, yet somehow they don't lose 76-0 or go years without winning on the far side of the Equator.' That was the general argument and, in many ways, it was a difficult one to rebut, and remains so to this day. I might say in response that while England – and, indeed, the other home nations, along with France – were making their summer tours at the end of a fragmented fixture list (club, Tests, club, Six Nations, club), the Beautiful South flew north with far more cohesion and continuity, having been together since the start of their international programmes five months previously. Had we set off for New Zealand in that summer of 2008, straight off the back of the Six Nations, we would have been tired but connected, rather than tired and detached. I might also point out that a restructured season with a solid block of international championship and tour rugby stretching from early April to mid-June was a central component of the Andrew Report. Oh well.

I took the view that for all the strife generated over many years, England's summer visits to the great rugby strongholds of the south continued to be of value. There were exceptions, of course: the 1998 'tour of hell' was ridiculous in terms of its itinerary, even before it became doubly ridiculous in terms of its personnel. Two Tests in New Zealand and another in Australia a few months after the 2003 World Cup victory, when the whole of English rugby needed to draw breath and get to grips with some legacy planning, was probably not the brightest idea. As for 2007 in South Africa, we should never have gone within a gazillion miles of the place, even though it signalled the international resurrections of Messrs Gomarsall and Regan. (Nowadays, the major European sides prepare for World Cup tournaments by staying at home and playing warm-up matches with a specific purpose in mind, rather than travelling to the back end of beyond at someone else's behest and getting thumped for their trouble.)

I still believe that tours in non-World Cup years have plenty going for them. England squared a series with the Wallabies in 2010 and could easily have won it; we gave an encouraging account of ourselves in adversity in South Africa two years later, drawing the last game with the Boks in Port Elizabeth; we probably shaded the first Test and a half against New Zealand in 2014, even though we ended up on the wrong end of a 3-0 blackwash. In the end, it comes down to context: a team at the right stage of development can draw enormous benefits from exposure to the strongest opposition in the furthest-flung corners of the rugby landscape, and the fact that England have been far more competitive in the summers since 2008 than they were previously is proof that our game at international level rests on stronger foundations.

But as we looked ahead to the autumn series in 2008, those foundations were still relatively weak. Martin Johnson was fully on board in his manager-coach role, the torments of Christchurch were behind us, and I was back in the day job, but I knew there were mantraps on the road ahead of us. The squad was reshaped – almost a dozen of the players involved in the New Zealand Tests in June were no longer under immediate consideration – and there seemed to be more strength to it: there were a couple of major-league tight forwards back in the mix in Phil Vickery and Simon Shaw; there was a World Cup winner in Josh Lewsey and a 2007 finalist in Paul Sackey among the three-quarters; there were some of the brightest young talents to emerge in a generation, from James Simpson-Daniel and Danny Cipriani among the twinkle-toed brigade, to Dylan Hartley at the sharp end. What happened? We lost three of our four matches, two of them – against, you guessed it, the All Blacks and the Springboks – by record margins. It was another painful experience, sitting there thinking 'bloody hell, we're a mile off.' There were times during the 42-6 defeat at the hands of the Boks, who claimed all five of the tries scored that day, when my mind flashed back to that eye-watering, stomach-churning afternoon with the Under-19s in Belfast the previous year. Men against boys, again. That South African side was the best they had produced in many a long year – Bryan Habana, John Smit, Bakkies Botha, Victor Matfield, Schalk Burger – but it was only 12 months since we had pushed those same players every inch of the way in the World Cup final. Whatever had happened since the turn of the year – and plenty had happened, one way or another – it had not been great. I was almost as keen to see the end of 2008 as I had been to see the back of Christchurch.

10
BOTTOM OF THE BARREL

JUST WHEN YOU think you've seen all there is to see; just when you assume that an upturn in fortunes is inevitable because there is no deeper place for the downturn to go; just when you start believing that the bottom line has been drawn as low as it is humanly possible for anyone to draw anything – that's when you're guilty of kidding yourself. The 16 months or so between Francis Baron's exit as Rugby Football Union chief executive and the departure of Martyn Thomas from the corridors and committee rooms of Twickenham were perhaps the most chaotic in the history of top-level sport in England. During that spell, the organisation became embroiled in what amounted to a civil war and very nearly collapsed under the strain. Things reached such a pretty pass that the government waded in with demands for a root-and-branch review of the entire set-up. Not rugby's government, the International Board, but the Westminster version, MPs, ministers and all. You know you're in trouble when the tribunes of the people start hinting at a withdrawal of public money.

This chaos was not confined to the RFU's own political class. It spread everywhere, enveloping the professional staff and, ultimately, the England team itself, ending in appropriate fashion with the tragi-comic World Cup campaign in New Zealand in 2011. If the events surrounding that tournament and its immediate aftermath sent us hurtling towards rock bottom, the union has been on the road to recovery in the years since. While such roads rarely run straight and true (our own global tournament, staged in 2015, was not so much a bump as a crater), I am confident that Twickenham now has the right management and governance structure to fulfil its potential as one of the great powerhouses, if not the greatest, in the game. Why? Because the RFU is no longer in the hands of people who are self-evidently unfit for purpose – many of whom find their way into positions of power despite having none of the requisite qualifications of playing, coaching or managing rugby at the elite level.

I knew full well that Martyn Thomas, the RFU chairman at the start of this lamentable episode, was no supporter of mine: he had been in Clive Woodward's camp when the elite rugby director role was up for grabs in the summer of 2006 – one of his principal cheerleaders, indeed – and as far as I could see, he had not changed his mind on the subject. But at this stage at least, I had better things to think about. There was a new agreement in place with the clubs, but there was still a mountain of work to be done at both ends of the representative game in England and, while it was never a part of my role to be a hands-on figure in the senior international camp, it was very definitely within my remit to offer any support that might be needed. As recent results suggested that the team needed all the support it could get, I spent a good deal of time talking things through with Martin Johnson and trying to strengthen the links between the age-group sides and the shop-window team.

Whether Martin Johnson was a good manager-coach or an inferior one, and views were divided on the subject from the moment he walked in the door to the moment he walked back out again, this much was indisputable: he was a realist to the core. He knew rugby backwards and he accepted, I think, that however many coaches of the past had conned themselves into believing otherwise, it was simply not possible to turn a sow's ear into a silk purse. Essentially, he was unconnable. He understood the limitations of the players available to him and recognised that while it was always possible to prepare and select a little better as time went on, the key to England's long-term success, which could be simply defined as the national team's ability to win big matches regularly rather than squeeze out the odd result now and again, lay with the next generation. Until the age-group reforms we had put in place bore fruit, all he could hope to do was tough it out.

The word 'supercoach' is now a part of the modern rugby vocabulary, but I don't think Martin believed in this mumbo-jumbo any more than I did at the time. (I believe in it even less now, despite the impact Eddie Jones has made on the England team since his appointment after the failed World Cup campaign in 2015.) The best coaches in the world, including Eddie, have had spells in their careers where results have made them look distinctly average. Would England have been a whole lot better around 2009 if someone other than Martin had been in charge? I have my doubts.

Brian Ashton might have been right in thinking that he could squeeze another 10 per cent out of some of his players: those who considered themselves to be his fellow travellers, like Danny Cipriani, often showed signs of growth. But was he a 'supercoach'? Does such a being exist? Please. Clive was lauded from the rooftops for winning the World Cup in 2003, and rightly so, but what kind of 'supercoach' was he when, with Johnno in retirement and a bunch of fading England personnel in his squad, he took the British and Irish Lions to New Zealand two years later? He had more money, more back-room staff and more players at his disposal than anyone in Lions history, yet the All Blacks won the Test series almost without breaking sweat. Was he a 'supercoach' when we lost three Grand Slams on the bounce? The notion that he would have made a huge difference to England in late 2008 and 2009 doesn't stack up: it made no more sense than the hopelessly romantic, misty-eyed reasoning of Martyn Thomas in recruiting Johnno to the England job, which seemed to be on the infantile basis that a great player was automatically equipped to be great at everything else in rugby. Johnno knew, as I did, that the crucial factors in England's future success were the underlying ones.

Interestingly enough, it was in this field that Martin's eventual successor, Stuart Lancaster, first came to prominence at Twickenham as head of elite player development. This was a 'does was it says on the tin' role, incorporating, as a challenging extra, the coaching of our second-string Saxons team. Conor O'Shea, the former Ireland full-back and London Irish coach who is now top dog in Italian rugby, had been overseeing the academies since 2005, but he decided to leave the RFU after being offered the national directorship of the English Institute of Sport. We were reluctant to see him go and worked hard to change his mind: he'd been highly effective in shaping the early talent-spotting programmes that would eventually lead to a fistful of Under-20s world titles. But his mind was made up, and we were left with a hole that badly needed filling.

Stuart offered a solution. He was well organised, on top of his brief, full of ideas and extremely persuasive in his presentation, so we brought him to Twickenham from the Leeds club (not that Leeds were at all happy about it), confident in the expectation that he would build on Conor's work. Which he did. He was as clear as I was that until we created an environment in which our age-group players could not only compete with their peers from New Zealand and Australia and South Africa but actually beat them, their chances of winning games against the same people a few years down the line at full Test level would always be limited. Better Under-18s meant better Under-20s. The trick then was to give those players regular exposure to Premiership rugby before feeding them into the senior England set-up as and when their performances justified it.

Historically speaking, we've been pretty poor at this in England. Not just in rugby, but across our major sports. If the people in charge at Twickenham spent far too long being overly tentative about fast-tracking the brightest young talent into the national team while the Australians were selecting such great players in the making as Tim Horan and John Eales and Phil Kearns in their teens or very early 20s, it has rarely been much different in football. One startling statistic in the round-ball game is that German players aged 21 and under play five times as many matches in the Bundesliga as their English equivalents do in the Premier League. Five times as many! It tells you all you need to know. I was lucky in my on-field career: I played a good brand of rugby at Cambridge University and had my first taste of top-end club rugby with Nottingham as soon as I graduated. Would I have been the player I became had I spent season after season on the bench, or been farmed out to a team playing substandard rugby that didn't even begin to stretch me? The answer is obvious. As a rugby nation we are now much better at producing youngsters capable of playing like grown-ups, which is just as well. As I've already made clear, the one certainty about sport at international level is that you can't buy your way out of trouble on the personnel front. If you're short of class at scrum-half or No. 8, you can't simply ship one in from Brisbane or Bloemfontein or Bordeaux.

With so much Twickenham business still in a state of flux I didn't exactly head for the beach on my return from the troublesome New Zealand tour in 2008 and, as the autumn international programme went from bad to worse, there was no escaping the build-up of pressure, from inside the RFU as well as from the public outside it. The fact that the media continued to misunderstand my role, or even appeared to deliberately misrepresent it, was bad enough: with every painful defeat, someone somewhere would be trotting out the 'Andrew must go' line. What made my life considerably more difficult was a growing feeling that my position was being undermined from within. It made for an interesting dynamic, to say the least.

Had Martyn Thomas been willing to stand eyeball to eyeball and say, 'Rob, thanks very much, you've done a good job for the union but I'm moving you on because I want to bring someone else in', I would at least have known where I stood. But that wasn't Martyn's style I even had to learn from a journalist that Clive had been invited to brief a group of RFU council members on what he would do if he were in my job. On one level, this was quite funny: the humour might have been as black as pitch, but I felt able to laugh nevertheless. On another level, it was the most extraordinary leadership-management-governance situation I'd ever experienced. Really, quite unbelievable.

While Francis was in place as chief executive, I felt I could rely on his support. I also knew I had the strong backing of some highly influential council members. But Francis would not be long for the world of the RFU: after 12 years of momentous upheaval, he left the union in 2010 after losing one fight too many with those committee die-hards who still laboured under the delusion that 1995 had happened everywhere but in England. There had been tension between the union's modern-minded executive and its more traditionally driven membership from the moment Francis took over at Twickenham, and he was no stranger to attempted coups by his enemies. Even those who admired his work in turning things round financially and putting the union on a secure commercial footing could see that he wasn't everyone's cup of tea, but while he took precious few prisoners during his time in TW2, he at least made himself clear. You knew he was being straight with you, even if it hurt.

Francis was what you might call a 'proper' CEO and, as such, he broke new ground at Twickenham in all sorts of ways – not least in his willingness to stand up to the old-timers on the council by insisting that the problems of modern rugby required modern solutions. There were personality clashes and policy disagreements with a number of people, including Martyn: the question of who would be driving preparations for the 2015 World Cup, which had been awarded to England during the summer of 2009, was a particularly vexed subject. No one on the Twickenham staff was under any illusion about the differences of opinion between them, so it did not come as a complete surprise when we heard that Francis was leaving the union and that the search was on for a new chief executive.

I was bang in the firing line now and I knew it. In fact, if Martyn had come to me straight and said, 'Time to go Rob, here's your pay-off', I couldn't have done much about it. There was nothing he could pin on me – the club agreement had been secured and activated; the age-group project was bearing early fruit; in all but the most theoretical sense, the England team was someone else's responsibility – but the way I saw it, there was not much in the way of logic at work. If Martyn had a rationale, it had a lot to do with restoring Clive to Twickenham in a senior capacity.

The Woodward supporters in the media were certainly back in full voice – shortly after the 2010 Six Nations, which had not gone terribly well, there was a 'special report' in the *Daily Mail* stating that Clive was prepared to 'answer the nation's call as and when Twickenham ask him to return to rescue English rugby', and that as senior RFU figures were reluctant to 'sacrifice' Martin Johnson, the man in jeopardy was me. Apparently, my 'star had fallen' during the Six Nations. The decibel level only increased when John Steele was appointed as the new CEO a few weeks later. A lot of journalists were under the firm impression that John would lay the ground for Clive's return, and it didn't seem to me that they were making it up. I had known John for years. I'd played against him during his lengthy stint as Northampton's outside-half, and renewed acquaintances in the Premiership when he spent a couple of seasons as head coach at Franklin's Gardens. He was coming to the RFU from a similar role with UK Sport and was bound to have big ideas on reshaping the top end of the organisation. What would this mean for me? I was keen to find out.

This much was certain: I intended to fight my corner. I knew from the moment I declared my candidacy for the elite rugby director's job in 2006 that I was putting myself in interesting territory, although it is fair to say that I didn't realise at the time just how interesting it would turn out to be. Now, four years into the job, I felt a surge of the old 'bring it on' spirit. I don't know where it comes from, but I can be seriously stubborn when the shit hits the fan, as the saying goes. And the fan was gathering speed. I remember thinking to myself: 'Okay, so Francis has gone and John is coming in and Clive may be on his way. You know what? This is quite entertaining.'

I will never forget our first meeting on RFU business, which took place in London at the One Aldwych hotel, not far from Waterloo Bridge. It was mid-July. I had just returned from England's four-match tour of Australia, where we'd won a Test in Wallaby country for only the third time in our history (we'd also played a game against the New Zealand Maori in Napier, which hadn't gone quite so well). John had yet to come to us from UK Sport, but he was gathering his thoughts and testing the water ahead of his arrival at Twickenham and the immediate instigation of a management review. We discussed the tour and, when he asked me how I was feeling about things in general, I told him that I believed the England team was making progress, albeit a little slowly, and that the age-grade side of our rugby was moving forward rather more quickly. 'Next year's World Cup will be a challenge,' I said. 'It's in New Zealand, after all. But we're getting there.' And that was it. As I walked back across the bridge to catch the train back to Twickenham, my phone rang. It was my younger brother Richard. He told me that our father Raymond, who had been suffering from cancer for some time and had just been through major surgery, had died that lunchtime. My head was all over the place. I'd been fearing the worst because Dad had been so ill, but the news still knocked me sideways. As I stood there, looking down at the Thames, there were no positive thoughts in my mind. If the summer of 2008 had been god-awful, the summer of 2010 was turning out to be a whole lot worse.

John started his new job a few weeks later and set in train his 100-day structural review with the help of a team of external consultants, who no doubt cost the RFU plenty. Quite why people appointed to take charge of an organisation immediately hand over this kind of work to others, I have no idea. I consider it to be something of a charade, but it happens all the time. I didn't have a good feeling about the process right from the start, but I didn't expect to learn about the big idea emerging from the review in the way I did. I was back home in Yorkshire for Christmas and New Year, spending time with my mum and thinking over one or two conversations I'd had with John, who I felt was being more than a little coy about any role I might have in his revamped RFU management team. I knew there was about to be a meeting, supposedly secret, where he would present his new structure to the board, but to the acute embarrassment of the governing body, details of John's proposals were leaked to *The Times* and reported by the journalist Mark Souster, who would go on to make something of a habit of this kind of thing. It seemed John was planning to split the rugby department into three directorates: performance, operations, development. As there were obvious ramifications in terms of the job I'd been doing for almost four and a half years, I phoned John from the family home without delay. 'Interesting stuff in *The Times*,' I said. 'What role do I have going forward? Is this constructive dismissal? No one's spoken to me about how this is going to play out.' John denied being the source, saying he didn't know where the leak had come from.

To my mind, two of the three 'directorates' already existed: 'development' was simply another name for the grass-roots or 'community' game, which was then being run by Andrew Scoular. As for 'performance', it was a term that covered a substantial chunk of my existing responsibilities. The new element was 'operations', which appeared to relate directly to what remained of those existing responsibilities. A part of me wondered whether this signalled the end of my spell with the RFU – whether I should explore the 'constructive dismissal' option and get the hell out of there. Yet I was intrigued by the operations role, partly because I felt it was important to build on the constructive relationship I'd helped create with the Premiership clubs, and partly because it would free me from a high-profile task – the overseeing of the England national team – that I now considered virtually impossible to execute in a satisfactory way. In effect, I now subscribed to the view expressed by the South African coach Nick Mallett when he was linked with the England job in the mid-2000s: namely, that a head coach must be in complete control of his team, stand or fall on his results and be answerable to the chief executive alone rather than to a third party occupying some ill-defined space in the middle ground. To me, it was not actually possible to carry out the job of performance director. Even had I been in the running for the post, which I knew I wasn't, I would have wanted no part of it.

There was a straight choice, then: to cut and run (roughly translated as 'give me a fat cheque and I'll go'), or to dig in my heels (roughly translated as 'I've busted my arse for four years trying to make sense of this place; we're on to something at long last and I'm buggered if I'm going to let someone waltz in and take the credit'). In the end, I decided there was enough of interest in the 'operations' job description to justify chucking in an application, and I was duly appointed in February 2011, seven months before the World Cup in All Black territory. The 'development' role was also filled, with Steve Grainger effectively replacing Scoular, who, in the new chief executive's own 'day of the long knives', had joined a handful of other Twickenham executives in disappearing through the exit, including the head of media Richard Prescott and the competitions director Terry Burwell. That left the headline position, with the Woodward supporters in the press banging their drums so loudly that you would have struggled to hear Led Zeppelin above the din. There were some interesting alternatives in the mix, too: Jake White – yes, the same Jake White who had been locked together with Andy Robinson in that morbid 'him or me' episode in the autumn of 2006 – was being talked up as a candidate, as was a certain Eddie Jones. I could barely wait for the outcome of this contest, not least because I was effectively combining my new role with my old one on an interim basis. Multiple plate-spinning had become something of a speciality of mine, but the novelty had been wearing thin for quite a while.

And then, all hell broke loose. After weeks of flagging up the performance directorship as a kingpin position, with Martin Johnson reporting to the new man on a basis significantly more formal than the one on which he had been reporting to me, John suddenly issued a new job description, stripped of responsibility for the senior England team. I was pretty close to this stuff, but to this day I have no clear idea as to the motives behind the change of tack. What I do know is that by diluting the role in the way he did, John set in train a meltdown in the RFU that would not be reversed until the end of the year, by which time England had departed the World Cup in a state that bore a depressingly close resemblance to disgrace, and the governing body had finally shed its 19th-century skin and set itself on the road to modernity.

I happened to agree with the decision to change the job description, because I felt the original one was fundamentally flawed. I knew, because I had long been a part of something similar. Looking back, I think I suspected there were problems with the management structure when I agreed to take on the elite rugby directorship in 2006, but at that point, Francis Baron was uncomfortable with the idea of line-managing the head coach. It is an incredibly difficult area at the best of times, shrouded in shades of grey. So much depends on the professional characteristics of the people involved: if the chief executive is essentially a businessman, as Francis was, he is unlikely to be as comfortable in his dealings with an England head coach as a CEO with a deeply rooted rugby background, like John Steele. To my mind, it is more productive and more sustainable for a performance director to concentrate on ensuring that the component parts of the system are properly connected – to take care of the politics, fight the financial battles, drive the academies – and leave team matters to the man in the tracksuit. If you give the coach the players and resources to do the job and the job isn't done, that's his problem. If you don't deliver on those fronts and the England team suffers as a consequence, the problem is yours.

Did John arrive at the same conclusion? Did he decide that Clive or some other big name from the international coaching community would make it impossible for Martin Johnson to carry out his duties – be too much of a fox in the henhouse? Did Martin have reservations himself? If so, did he make those reservations known? Even to those of us spending our working weeks in the Twickenham office complex, along with the main protagonists, this was parallel universe stuff. There had been plenty of times over my quarter of a century in top-level rugby when I questioned my own senses. Now, it was happening on a daily basis. 'I don't believe what I'm seeing' was one constant refrain. 'Did I hear that right or have I just imagined it?' was another. For a union with an in-built resistance to change – a resistance that had been most manifest in 1995, that year when the ostrich-like behaviour of the RFU hierarchy had resulted in the game slipping from their control – it was amazing how quickly things happened between the point in mid-March when Clive was widely reported to be the only serious contender for the performance directorship, and the point in early June when John Steele's brief reign as CEO ended in a forest fire of negative publicity.

At a hastily arranged board meeting that was by no means fully attended, it was decided on a split decision that the revamped job description should be re-revamped without further ado, with senior England team responsibilities restored to their place at the heart of the remit. John's plan to water down the performance directorship had not survived its first sight of the enemy. A week later, it emerged that Clive had ruled himself out of the running for the job – not that he had ever publicly ruled himself in – and would be staying in post at the British Olympic Association. Five days after that, the board reaffirmed their full support for John while, almost in the same breath, asking Peter Baines, the chair of the governance committee, to 'review' the CEO's handling of the recruitment process. (A cynic might say that these two steps were somewhat contradictory.) The inevitable followed a few days later, after another board meeting, this an emergency one beginning in mid-evening, continuing into the dead of night, and ending with John's departure. I was not among those invited to attend: I went to my fair share of those things, but on this occasion only voting members were summoned, along with Karena Vleck, a specialist sports lawyer and head of the union's legal team. Various criticisms were levelled at John – some stuff about strained relations with sponsors; other moans and groans about staff morale – but it was his perceived failure over the performance directorship that made him vulnerable.

So now we were without a CEO as well as a performance director. This was at least as bad as 2007 and probably worse, and there was power to add in the shape of yet another investigation, yet another resumption of boardroom hostilities and a catastrophic World Cup campaign with an aftermath to match. What about the actual rugby, I hear you ask? Just for once, this was the least of our problems – until, that is, we reached New Zealand for the big tournament. Our squad was still too reminiscent of a curate's egg for comfort – strong in the back row, for instance, but not so strong at centre – and there were only half a dozen or so hardened Test players with 50-plus caps in the shake-up for places. But at the other end of the experience spectrum, good things were happening: the flanker Tom Wood and the loose-head prop Alex Corbisiero were on the threshold of their productive Test careers; the wing Chris Ashton, the scrum-half Ben Youngs, the tight-head prop Dan Cole and the lock Courtney Lawes were also coming through the system. When we opened our Six Nations account with a comprehensive victory over Wales in a floodlit game in Cardiff – we were 23-9 up by the hour mark – and then put the best part of 60 points past the Italians at Twickenham, with Ashton hogging the limelight courtesy of the first four-try championship haul by an England player since the outbreak of the Great War in 1914, there were legitimate reasons for optimism. If the next two chapters, home victories over France and Scotland, were written more in prose than in poetry, Jonny Wilkinson provided a flourish: benched in favour of Toby Flood throughout the competition, he became the most prodigious scorer in the history of Test rugby when he kicked a second-half penalty against the

French, and then nailed a last-minute kick to take us out of range of a particularly cussed band of Scots.

After all we'd been through in Six Nations terms since the exceptional Grand Slam display in Dublin back in 2003, it felt good to be back on top of the pile, and while our best-laid plans for a repeat performance came unstuck against Brian O'Driscoll and his friends in the final match, the title was ours. Mark Evans, my old acquaintance from the days of the Andrew Report, liked to describe drawn rugby matches as 'the equivalent of kissing your sister' and there was something of the same feeling when the England players, some of them slightly the worse for wear after waiting for the conclusion of the France–Wales game that evening, took possession of the trophy off the back of a defeat. But all things considered, we would happily have settled for solid silverware at the start of the campaign. With no summer tour to concern us – June trips in World Cup years had bitten the dust after the lunacy of our visit to South Africa in 2007 – I felt Martin at least had some momentum behind him.

But of course, there was always the problem of Martyn Thomas to set against the progress of Martin Johnson. John Steele's screeching U-turn over the performance directorship, followed by the response of the Thomas cabal and the effect on the top end of the Twickenham management structure, put the board on a collision course with the council. Almost immediately, Judge Jeff Blackett was back on the disciplinary beat, conducting a council-commissioned investigation into the Steele affair and all that surrounded it. He pledged to deliver a 'warts and all' report and, to this end, he went through a library-load of documents, gathered together dozens of written submissions and interviewed all the main players.

I was among those interviewees and I answered a range of questions to the best of my ability. I didn't consider myself to be a central figure – I'd spent much of the year getting to grips with my new operations role, filling the performance director vacuum as best I could and generally keeping my head down – but I was happy to co-operate with Jeff in his attempts to establish exactly what had gone on. His report was fiercely critical of Thomas as an individual and, with the odd exception, the board in general – criticism that immediately resulted in threats of legal action. The majority of council members were determined that Thomas should stand down as chairman, which he eventually did. What I did not foresee, in my wildest nightmare, was him staying on as acting CEO – a role he had taken on in the immediate aftermath of the Steele business. This was beyond satire, beyond bizarre. 'You really will not begin to believe what I'm about to tell you,' I said to my team in the office. 'This trumps everything by a million miles.' For want of a more extravagant word, I was incredulous. Here we have an organisation in pieces, with old friends and colleagues clawing out each other's eyes to the sound of mocking laughter from the English sporting public, and I suddenly find that the man who I felt had me in his sights as chairman has somehow found his way into an even more powerful position and is now my immediate boss. Could this really be possible? Apparently so.

We were now in July. In five weeks, we would be playing Wales at Twickenham in a World Cup warm-up game. In eight weeks, the squad would be flying to Auckland for the main event. Martin Johnson had enough on his plate. 'Just keep me away from the politics, Rob,' he said to me. 'Build a partition wall and leave me on the far side of it.' Which was fine. But there was no partition separating me from what was going on behind the scenes. 'If Martyn Thomas wants a fight,' I said to myself, 'I'll give him a fight.'

As mentioned earlier, there comes a point when I relish the thought of a toe-to-toe scrap. I loved it when I was playing, when all that Rob Andrew versus Stuart Barnes stuff was flying around, and I was more than prepared to dig in now. On the field, off the field – the more someone disses me, the more I'm likely to react by doubling and tripling my efforts to prove them wrong. Especially when I *know* they're wrong. I'm not saying I hadn't made mistakes during my time at the RFU, any more than I was error-free as an England outside-half. Some of the situations surrounding the national coaches had been very messy, for instance; things should have been done differently and I was partially to blame for the fact that they weren't. But when I analysed my own performance during those wild summer weeks at Twickenham, I felt I'd done a decent job in most respects. There was a club agreement in place, and for more than three years there had not been even the slightest suggestion of a club–country spat; the age-group project was up and running and gathering speed; the senior England team had, for all its travails, brought home a Six Nations title.

It was at this stage that Martyn, sitting in the CEO's chair, started re-examining the management structure for the umpteenth time. Mysteriously, the performance director's role had disappeared into thin air. The buzz phrase now was 'professional rugby director', PRD. I upped the ante by asking my old friend Geoffrey Hamilton-Fairley to pay Martyn a visit as a means of finding out, directly, what the hell had been going on, while at the same time getting some kind of handle on what was likely to happen next. It had the desired effect, for the outcome of the meeting was revealing. Much to my amusement, the name in the frame for the new PRD job was mine. Effectively, I was being reinstated in my original role, swapping the word 'elite' for the word 'professional' but otherwise carrying on in the same old way. The one problem with this was that Martyn hadn't signed the confirmation letter. I knew the letter existed because the union's human resources people told me it did, but without the signature of the CEO, the deal wasn't worth the paper it was written on. It was for this reason that I missed England's first game of the World Cup, against Argentina in Dunedin. I stayed behind because a Twickenham staff meeting had been called for the Monday after that match and Martyn was scheduled to announce the new leadership structure. I wanted to hear him, with my own ears, say my name. It was only then that I boarded a flight to New Zealand.

There was a point where I nearly didn't go at all: my new job still hadn't been signed off and as my existing responsibilities as operations director did not cover the England team, I was tempted to stick two fingers up to the union and watch the World Cup from the safety of my lounge on the basis that if things went pear-shaped, I would be well out of it. And as news was already beginning to emerge about certain England players winding down from the narrow victory over the Pumas by going 'on the lash' in the South Island adventure sports resort of Queenstown and making complete fools of themselves in a late-night bar, there did not appear to be much of an 'if' about it. All the same, I listened to my better angel and made the trip.

By the time I arrived, after a journey of 30 hours or more, the team were back in Dunedin ahead of the pool fixture with Georgia – not a game we were likely to lose, but an important one nonetheless in terms of team-building and statement-making. The first thing I saw on entering the hotel was Martin Johnson and Tom Stokes, the team operations manager, in earnest conversation in a corner of the lobby. It was one of those 'Oh Christ, what's gone wrong?' moments: a sixth sense told me that this was not good. I was tired, I hadn't even checked in, and already I felt uneasy.

'We have a problem,' Tom said as I approached, dragging my bags behind me.

'Go on.'

'There's a story about to break about a hotel maid claiming she's been trapped in a room and harassed by some of our players.' I think I knew there and then that this would be the World Cup from hell, that I would spend the next month of my life fire-fighting for all I was worth.

On reflection, this should not have come as any great surprise. There had been an undercurrent of trouble even before the party left England and, not for the first time in the recent history of the national team, it concerned money. The players informed the union that they were dissatisfied with the financial arrangements in place for the World Cup. I considered this to be outrageous behaviour on their part, not because I was against players pushing for improved terms and conditions – heaven knows, I'd done the same myself on more than one occasion – but because the figures had been agreed as part of a four-year deal between the RFU, the players' union and the members of the England squad. It was all there in the contract, in black and white: match payments, bonuses, you name it.

I'd been involved in the original negotiations, as had the players' association boss Damian Hopley, and as far as I was concerned, it was binding. The players saw things differently and they tried to hold the union to ransom. When we held a World Cup departure dinner at Twickenham, which was a very grand affair, attended by all our major sponsors and commercial partners, they more or less refused to get off the team bus. It was totally unnecessary, utterly irresponsible and entirely without foundation.

Johnno was furious. Like me, he knew what it was to lock horns with the RFU over money: during his time as captain, he was one of those who had threatened strike action before a Twickenham Test against Argentina. But in that instance, contractual negotiations had reached stalemate. Here, everything had been signed and sealed and was in the process of being delivered. The incident left a sour taste in the mouth. Some of the senior figures in the England party had been led by Martin, had played alongside him in some of the greatest victories in the country's rugby history, yet they were prepared to let him down. He deserved their respect and received the opposite. How some of them can look him in the eye today, I have no idea.

The fallout from the Queenstown incident was toxic in the extreme, largely because Mike Tindall was one of the principal protagonists. Mike was a senior player, a World Cup winner and team leader – indeed, he had captained the side in the opening match in the absence of the injured Lewis Moody. What was more, he had just married Zara Phillips, the Queen's granddaughter, and was therefore an obvious target for chancers with camera phones seeking to make easy money from the tabloid press. After the tournament, he was dropped from the England set-up and heavily fined for a breach of the Elite Player Squad code of conduct. Mike was not best pleased with this outcome and accused the union of scapegoating him, but in truth, his behaviour had not been within driving distance of acceptable. As for the chambermaid business in Dunedin, newspaper reports of serious wrongdoing turned out to be wholly false. But the situation should not have arisen in the first place: it was messy and distracting; it ate up man hours that would have been better directed elsewhere. In strict legal terms, the problems we faced during the early stages of the tournament were not on the scale of the 2008 calamity, but the line between the interests of the RFU and the interests of the players was a difficult one to tread.

In 2008, we had been able to present a united front. At the World Cup, there was some damaging divergence. The whole tenor of the campaign was shockingly bad and, deep down, I think it really hurt Martin. I spent plenty of time meeting with players and agents and legal representatives; I had more overnight discussions with the RFU legal department back home than I care to remember. But it was far worse for Johnno, who saw it as his job to front up in public and was reluctant to let anyone else share the load. He seemed to be on his knees by the time we were knocked out of the tournament: a giant of a bloke, laid low by people he thought he could trust.

The peculiar thing was that, despite the substandard rugby we produced in the pool stage (the Argentina game was too close for comfort, and we could easily have lost to the Scots), a route to the final had opened up for us, just as it had in 2007. France in the last eight? Wales or Ireland in the last four? Here was a chance to pull something from the flames. But while French blood had been frozen in the veins before kick-off in our previous two World Cup knock-out meetings, it ran hot on quarter-final night in Auckland. They were barely more together than we were – looking back, it is possible that they were even more of a rabble – but there were some very good players in that side and two of them, the scrum-half Dimitri Yachvili and the No. 8 Imanol Harinordoquy, were absolutely in their pomp. We could not hold them in the first half, during which they built up a 16-point lead, and while we scored a couple of tries after the interval, there would be no way back. Disappointing? Of course. Was I sorry it was over? Probably not.

Between our departure from the tournament and our departure from the country, there seemed no obvious window during which any of the players could present us with another bucket-load of grief. Yet Manu Tuilagi, the Leicester centre, found one and promptly threw it open. After defeat by the French on the Saturday night, the management decided to have one final dinner together early on the Sunday evening, and duly gathered at a restaurant in the city's harbour area.

The events are etched upon my mind. I'm sitting opposite Johnno, who has Tom Stokes, the team operations manager, sitting next to him. The mood is an odd mix of the sombre and the darkly humorous: we're in a 'let's get out of this godforsaken place' frame of mind. After a couple of beers and a mouthful of food, Tom's phone starts ringing. 'Yes, it's Tom speaking . . . Slow down . . . what's happened? The police have . . . *What*? He's jumped into *where*? And he's been arrested? Where is he now? Right, let me just have a chat with Johnno.'

Looking directly across at our thoroughly beleaguered manager, I can see Johnno's face has turned the colour of a bottle of milk. 'That was Floody,' says Tom, referring to Toby Flood, whose form in the England midfield had been one of the few saving graces of the trip. 'Um . . . Manu has . . . um . . . been fished out of Auckland Harbour. It seems he jumped off a ferry intending to swim into shore and he's been nicked. Everyone's all right, but we need to get QC to sort him out.'

I'm flabbergasted, well and truly. I simply cannot process the 'what, how, why, where, when' aspects of this story, although it soon transpired that the players had sailed across the bay to Waiheke Island to enjoy a few drinks – why not end the trip as they had started it? – and events had unfolded from there. Manu had stated his intention to hop over the side on the way out, on the grounds that this was a traditional form of disembarkation in his native Samoa. It was not the brightest of ideas, given that he was on a bloody great ferry rather than a small fishing boat, and one or two players talked him out of it. On the way back, however, there was no stopping him. It was a case of, 'You're not doing it, Manu . . . Jeez, where's he gone?' It was the final ignominy. How in God's name, I remember thinking as the details emerged, are we going to explain this one to the outside world?

There would be plenty of opportunity for explanations, of course: a massive post-tournament review was underway almost the moment we touched down on English soil. I felt incredibly sorry for Martin, who could have ruled with an iron fist but chose not to because of the deep level of trust he had in the players. How could he have predicted that they would let him down so badly, have betrayed him so completely? I also felt for the rest of the back-room staff, who had worked around the clock to put out the fires started by the very people they were there to support – individuals like the former Sky Sports broadcaster Will Chignell, who had not long been in post as the RFU's communications director and, through no fault of his own, suddenly found himself under a degree of pressure he could not conceivably have anticipated. How some of the worst-behaved members of the squad could live with themselves, I could not begin to fathom. What I did know was that the politics had yet to be resolved.

Indeed, the situation was even more fractious than it had been during the summer, primarily because Martyn Thomas, in Auckland for the quarter-final, had loftily declared that his old mucker Fran Cotton would be conducting his own review into events at the World Cup. It was an astonishing move that flew directly in the face of the procedures we had painstakingly put in place under the auspices of the Professional Game Board, the joint union–club body set up to administer professional rugby in England under the agreement hammered out in 2007. It was also the act that ran down the curtain on the Thomas era.

Martyn had no authority – no damned right, to put it bluntly – to set up his own review: he was bypassing a system that had been established well before the tournament. Influential council members were fuming about it, as were the clubs: virtually everyone was fed up with the acting CEO's off-piste excursions, and newspaper reports indicated that there were more than a hundred clubs up and down the country who might push for the calling of a special meeting of the governing body unless he was removed. There was a fiery board meeting, during the course of which I had a blazing row with Martyn. And then he was gone. Less than a month after our return from New Zealand, he stepped down from all positions with the union.

The official review went ahead as planned. Premiership Rugby sought the views of those clubs who had provided players to the England squad; the players' union carried out the player interviews through their chief executive Damian Hopley; I sought the opinions of the RFU staff who had made the trip. Everything was fed into a single overarching document, full of incendiary detail and therefore strictly private. It stayed private only until it went public with almighty flashes of lightning and cracks of thunder. Quite how a hard copy of the report ended up in the hands of Mark Souster, our good friend from *The Times*, remains a mystery, although we all have our ideas about who was at the heart of the leak. What is very clear is that it created another unholy mess, with players slagging off coaches in the most personal terms. Perhaps the one relieving factor is that the leak of the review, a wickedly destructive act, did not force Martin Johnson to leave the England set-up. That decision had already been taken.

On his return to England, he spent a fair amount of time thinking through his options. By mid-November, we said to him: 'Which way do you want to play it, Johnno? Do you want to make a call yourself, or would you prefer to knock it back into our court?' In the end, it was his choice to walk away.

If I felt a sadness about Martin Johnson's departure, I felt very differently about Martyn Thomas's. All I felt – and still feel – in his regard is anger. In my view, the Rugby Football Union has 'mother of parliament' status in our game: it is one of the leading governing bodies in the whole of sport. As chairman, I felt he was woefully inadequate. We had our problems and disagreements as individuals, many of them profound, but my criticism goes way beyond the confines of a mere personality clash: it has far more to do with governance and standards and behaviour. Under his leadership, Twickenham was reduced to dust. It is not every day that a Minister of Sport writes to a sports body demanding that it gets its house in order, as the Tory MP Hugh Robertson did at the height of the internal fighting in 2011. For decades, the RFU had been run as a private club. The minister's letter was a timely reminder that, in the professional era, the public had a legitimate stake in its affairs. We had become a laughing stock. It was time to get serious.

11
GETTING IT RIGHT AND GETTING IT WRONG

IT HAD BEEN a long time coming, but the dinosaur-killing asteroid had finally struck Twickenham. New forms of life were emerging from the wreckage of the Rugby Football Union: suddenly, after a decade and a half of conflict which seemed to me to be rooted in the old guard's desperate fight for survival, there was a modernisation process underway. The first independent non-executives quickly materialised in the governing body's boardroom – Miles Templeman, an ex-director general of the Institute of Directors, and Andy Higginson, a former finance director of Tesco, were the trailblazers – and in the weeks either side of Christmas there was an air of regeneration about the place. We were still up against it with a Six Nations right around the corner and important positions to be filled, but at least there was a sense that we were all on the same side now. It might have seemed like a communion of damaged souls, but the overriding feeling was one of optimism.

Over the months following the appointment of Ian Ritchie as our new chief executive, and Stuart Lancaster as head coach, that optimism was widely considered to be justified. We would rediscover our bearings at international level far more quickly than had seemed conceivable during the terrible days of the World Cup aftermath: we would go close to winning the 2012 Six Nations title and give the Springboks a run for their money in South Africa, drawing a Test in Port Elizabeth; heaven forbid, we would even beat the All Blacks! There were those in the press who resolutely refused to support the low-profile, technocratic Stuart as successor to the stratospherically profiled Martin Johnson: even when presented with cast-iron evidence of progress, they refused to change their stance. But the vast majority of England rugby followers were happy to ally themselves with the Lancaster regime – until, that is, we fouled up at our own global tournament and became the first hosts in the history of the competition to miss out on a place in the knock-out stages of a World Cup.

If the story of the rise and fall is a bewildering one in some respects (only Stuart knows exactly why he took some of the decisions he did as the pressure built in 2015, particularly those involving the big-name rugby-league import Sam Burgess), I firmly believe that in important ways, he left the England rugby team in a better place than he found it. Everyone at Twickenham was affected by our 2015 World Cup failure: there was intense disappointment at the performance of the team, together with a feeling of genuine sadness that people who had poured so much of themselves into the project should have so little to show for their efforts. But there was no comparison with the 2011 experience. The RFU did not go into free-fall; it did not crash and burn. When Ian Ritchie flew to South Africa determined to lure Eddie Jones from Cape Town to London as Stuart's replacement, he did not have to throw himself on the Australian coach's mercy or give him the hard sell. Eddie would not have to start over. He would simply have to build on the foundations already in place. Among others, Stuart can take credit for that.

There was nothing solid about the union in the weeks leading up to Stuart's accession. When I left New Zealand at the end of that blighted 2011 tournament, all I could see was trouble. We had travelled home in dribs and drabs and I was one of the last people associated with the England party to land at Heathrow. Manu Tuilagi had just about dried out following his harbour-diving exploits, but English rugby was still wet through. If it wasn't completely washed up, it was drifting in on the tide. Would I have a part to play in turning things around? I was not in a position to say. Such was the extent of the havoc wreaked during the Martyn Thomas years, there was no guarantee that any of us would remain on the payroll. Would I survive the upheaval? It was hard to say.

Once Martyn had been removed, the RFU had to act fast – hardly something for which it was renowned but needs must. Ian Metcalfe, a fine player in his day for Cambridge University and Moseley, and the chairman of the Professional Game Board, was one of the men who came to prominence at that extraordinarily sensitive moment. Another was Stephen

Brown, who had joined the union as finance director that summer and now took on the additional role of acting CEO. For one reason or another, Martyn Thomas had never quite got round to signing off my supposed new job of professional rugby director, so now my future at Twickenham rested with the emergency hierarchy. Things would be decided at a board meeting on the last day of November.

I attended that meeting, but was asked to leave while my position was being discussed. I was outside for two hours, during which time I drank coffee with two of the people I trusted the most: Nathan Martin, a one-time Royal Marine who had first come to the union during Clive Woodward's spell as coach and was now head of operations, and Sarah Gilmore, my PA. They had been rock-like in their support and it was good to have them around, not least because I didn't have the faintest idea which way it would go. The upshot? I would be staying on, in the role to which Martyn had appointed me verbally but not confirmed me in contractually. It was my third job title in a year, but this time there would be no confusion over my duties vis-à-vis the national side. There would be no performance director, no buffer between the head coach and the chief executive. I would be overseeing pretty much everything else in the elite rugby department, but I would be spending my time behind the shop counter, not in the shop window.

All this was announced publicly at a packed press conference later that day, the entirety of which was screened live on television. It was quite an occasion: with the media in the mood for blood following the leaking of the confidential post-World Cup report, Ian and Stephen knew they were in a tough position. They handled it brilliantly. I cannot remember a more sure-footed performance from senior RFU figures in the face of hostile questioning. I also remember the conference for more personal reasons: the 'Rob's not going anywhere' message brought an immediate end to a media hammering, the like of which I had never previously experienced.

I'd taken my share of enemy fire as a player, especially from those who devoutly believed in Stuart Barnes as English rugby's long-awaited messiah in the No. 10 shirt, and had rarely been out of the crosshairs since. But this had been different. This had been really vicious and way off the scale. A lot of the press took the view that if Manu had gone overboard in the literal sense, I should be thrown overboard in the figurative one. The *Daily Mail* really went for me. One headline accused me of 'Shamelessly Holding On'; another described me as 'pompous' and claimed I was 'running for cover'. For the avoidance of doubt, the paper then published a giant mugshot of me all over the back page, next to the headline 'Go Now'. If they'd turned my head into a root vegetable, à la Graham Taylor, I wouldn't have been surprised. We always had copies of the major newspapers in the office, so I could hardly avoid reading what was being written about me, but I put on my Yorkshire-born-and-bred front, playing everything down the middle with the straightest of bats like some business-suited Geoffrey Boycott. It was brutal all the same, to the extent that some of the more reasonable rugby journalists certainly felt a degree of discomfort at the vitriol being peddled by their colleagues.

Yet when Ian and Stephen made their announcement at Twickenham, the whole thing stopped, dead in its tracks. There was barely another word published. It was as though the 'Get Andrew' mob thought to themselves: 'The story's gone. Right, on to the next one.'

I think most people on the front line of RFU affairs experienced a lifting of the spirits at this point. I know I did. There was certainly an air of celebration when the annual Christmas bash was held a couple of weeks later. It was a big do with most of the staff in attendance and, looking back on it now, there seemed to be a feeling widely shared, albeit subliminally, that there had been a cleansing of the organisation – that our place of work was somehow more wholesome, more civilised. I was profoundly aware of the support I'd received from so many people at Twickenham, a good number of whom were in the room that day, but that didn't stop me being taken completely by surprise when dozens of the party-goers suddenly donned Rob Andrew masks. I never found out who was behind the production, but it became the theme of the event after a few drinks. I ended up wearing one of the masks myself and, while it was entirely light-hearted, I also found it incredibly touching. These people had helped me through a pretty testing few months and I was grateful to them.

Even allowing for my mile-wide stubborn streak, I had not found the episode easy to endure. On more than one occasion during my time at the RFU, and particularly after the sacking of Brian Ashton as England coach and the appointment of Martin Johnson, Brian Moore, my old international colleague and comrade-in-arms, took to the national press to suggest that I should resign as a matter of principle. I cannot honestly say that the thought never occurred to me. But while it's simple to look at my position with the aid of hindsight, it's not so simple in real time. I made a mistake in not clarifying my responsibilities when I joined the RFU in 2006; I got things wrong during the Ashton affair; I probably shouldn't have accepted the team manager role in New Zealand in 2008. But at the end of 2011, armed with a new job description that I considered to be just about right in terms of its reach, I felt I still had something to give. We were advertising for a new chief executive, we were searching for a head coach; we had a team in pieces and a reputation to rebuild. I wanted to stay and fight.

Ian Ritchie was identified as our preferred candidate for CEO before Christmas and appointed in a short space of time. We could live with the fact that he couldn't start immediately. What we couldn't live with was the absence of anyone capable of running the England team with a Six Nations tournament just the other side of tomorrow. Brian Smith, the attack strategist at the World Cup, had followed Martin Johnson out of the door: he was fiercely loyal to Johnno, had been badly hurt by the revelations in the leaked documents, and had just about had his fill, so I wasn't remotely surprised by his decision. The other members of the back-room team – John Wells, Mike Ford, Dave Alred – were still with us, wondering what the future might hold. We owed it to them, to the English rugby public, and to ourselves, to act decisively.

Despite the state of things in the Test arena, there were a number of potential candidates and plenty of conversations. Nick Mallett immediately emerged as the ante-post favourite and there was the usual interest from Jake White, his fellow South African. These were the hardy annuals. Among the new names on the roster of 'possibles' were Dean Ryan, a familiar face from my days at Wasps and Newcastle; the former All Black wing John Kirwan, who had been a member of the great World Cup-winning side in 1987, and had gone on to coach both Italy and Japan; and another New Zealander in Wayne Smith, by common consent one of the sport's most formidable thinkers and an important contributor to his country's reclaiming of the global title a few weeks previously. With the exception of Jake, all these men were in London that December, working with the northern and southern hemisphere sides pieced together for the Heroes Rugby Challenge match at Twickenham. Some of them were extremely tempted by the England vacancy: in fact, I can't remember ever having more invitations for a 'quiet catch-up over coffee'.

I was in the thick of the coach selection process. Not to put too fine a point on it, I was pretty much on my own. This was autumn 2006 all over again, with the union saying: 'The clock is ticking, Rob. Crack on with it.' But if there were similarities between the two situations, there was one key difference: I didn't feel I was in the eye of a storm this time. I felt I was at the start of something rather than moving towards the end of something. And one of the people who best understood the requirements of the moment was Stuart Lancaster, who, like one or two of his better-known rivals, made it abundantly clear that he wanted the job. And with Nick Mallett, the front-runner as far as the media were concerned, playing things somewhat cagily by saying that he wanted to stay in South Africa for the time being and would not be available until the back end of the European season at the earliest, Stuart was in the right place at the right time.

He was far more ambitious than most people tended to realise on first meeting him: confident in his own ability, deeply interested in the latest trends in sports management and high-performance coaching, and more than happy to put in the hard yards – his capacity for work was entirely in keeping with his Cumbrian farming background. He also had good credentials. He had coached England's second-string Saxons team with a fair degree of success and, as I was the one who had brought him to Twickenham with the task of maximising the return from the academy programme we were funding so generously, I was as aware as anyone of how well he'd responded to the brief.

There was a good deal of pressure on the appointment, but not in terms of finding someone capable of delivering results immediately. The pressure was around timing. We had space in the calendar for a pre-Six Nations training camp, but we had nobody to run it. We didn't even have anyone in a position to decide where the camp would be held. When it came to tournament preparation in the dead of winter, Martin Johnson had taken a liking to Portugal. Were we going back there? Not a clue. For all I knew, we'd be spending a long weekend in Timbuktu. No one could say what was happening because no one was in place. And the match against Scotland at Murrayfield was less than two months away.

People might wonder how things could have reached such a pretty pass at the biggest, wealthiest union in world rugby. They might feel that somewhere along the line, we had allowed our coach succession planning to veer off track. And, on the face of it, they would have a point. But if it was as easy as all that, everyone would be doing it. The fact of the matter is that in England, the whole concept of succession planning is easier stated in theoretical terms than it is delivered in practical ones. The New Zealanders are really good at this stuff, as everyone else in world rugby knows to their cost, but the situation in which they find themselves is somewhat different to the English experience in the sense that, over there, the governing body controls its own market. The Kiwis routinely send their best coaches to overseas finishing schools, generally in the British Isles or France: John Mitchell was an assistant coach with England under Clive Woodward before taking on the All Blacks job; Graham Henry and Steve

Hansen both coached Wales at Test level; Wayne Smith learned a thing or two about northern hemisphere rugby during a productive spell at Northampton; Joe Schmidt and Vern Cotter coached at Clermont Auvergne before taking over Ireland and Scotland respectively, and will no doubt be in the frame when the New Zealand job next becomes available. The Kiwis show no sign of stopping, either. Todd Blackadder is embedded at Bath; Dave Rennie has agreed terms at Glasgow. Oh, almost forgot: there's been a bloke by the name of Gatland hanging around in this neck of the woods, too. At what point will we start using New Zealand as an educational hothouse for our own coaching talent? That'll be never. It does not serve the purposes of the All Black hierarchy to give foreign coaches access to their state rugby secrets, so they don't make jobs available to them.

This was one of the big beefs of Kevin Bowring, the Welshman who came to Twickenham in the early 2000s as head of professional coach development and spent 14 years trying to address the problem created by the fact that, unlike their fellow custodians in All Black country, the RFU exerted precisely no control over the domestic rugby market. In a system populated by Premiership clubs run as private businesses, there is no production line. Bath need a new coach? In comes Blackadder, all the way from the South Island of NZ. Worcester require some hard-headed realism as they fight for their top-flight survival? Who better than Gary Gold, a South African with a record of delivering backs-against-the-wall victories? Gloucester have a parting of the ways with Laurie Fisher, their Australian game-planner? Johan Ackermann is the next in line. That's Ackermann from Benoni in the East Rand, not Ackermann from Bengrove, east of Tewkesbury.

I had a lot of time for Kevin: to my mind he did a top job in seriously trying circumstances. But in the end, the realities of the situation held him back. There was no natural English successor to Martin Johnson in 2011, and we had not moved an inch in this direction by the time Stuart left the RFU after the 2015 World Cup. And I would not put so much as a penny of my savings on there being one when Eddie Jones, the first overseas coach to run the England machine, calls it a day after the next tournament in Japan. Unless things change in a way I simply cannot foresee, the union will have to continue making decisions in the moment, just as we did with Brian Ashton and Johnno and, yes, with Stuart and Eddie.

Stuart made it easier for us because we were in 'new broom' mode. We knew that the English rugby public had been pretty disgusted by the national team's behaviour at the World Cup in New Zealand, and that support would drain away unless we took action to 'drain the swamp', as President Trump might put it. Stuart was big on values, big on discipline, big on respect for the shirt. He came across as something of a puritan, and that was no bad thing, given the state we had got ourselves into as a Test team. Indeed, he knew as much as anyone about the condition of the national side, for he had been in New Zealand as the World Cup campaign fell apart, and so had been on the distribution list for the review document. (The official list, that is, not the unofficial one created by *The Times*.) We had travelled together to the quarter-final against France and watched the game side by side. I remember saying to him on the way to Eden Park that night: 'If we lose this one, what follows will be like nothing you've ever seen in your life. We might just about get out of this country intact if we make it to the semis. If we don't, the balloon will go up.' I was correct about all of that.

As time went on, it seemed to me that Stuart overplayed his hand on the 'cultural reconnection' side of things: there was so much emphasis placed on tub-thumping patriotism that the team's focus became blurred. But in the early days he struck precisely the right note in reminding a heavily revamped Test squad of their responsibilities to the wider rugby public as well as to each other. He also performed well in the crucial area of selection, both in terms of the playing squad and the coaching panel. Many of those guilty of unprofessional behaviour at the World Cup had been senior figures in the group and they quickly found that their Test careers were over. Those with their futures ahead of them rather than behind them were read the riot act and told in no uncertain terms that they were drinking in the Last Chance Saloon (not that Stuart was especially keen on them drinking anywhere). He knew virtually everything there was to know about the richness of the talent emerging from the Under-20s, but with the exception of the midfielder Owen Farrell, who was playing regular Premiership rugby for Saracens at an unusually young age, most of those individuals were still a little undercooked. So he built his first Six Nations party around people he knew and trusted, some of them a long way short of being household names but all of them made of the right stuff when it came to character and commitment. If he knew that a few of those drafted into the squad would win only a handful of caps – that they would quickly make way for the likes of Jonathan Joseph, George Ford, Joe Launchbury and Billy Vunipola – he also knew that they would squeeze every last drop of value from their moment in the sun.

At the same time, he made big moves on the coaching front. Stuart wanted to keep things tight: a three-man unit and no more. He had worked with Graham Rowntree in the past and had no qualms about keeping him on the RFU payroll and promoting him from scrum technician to forwards coach. He had also worked with Andy Farrell, father of Owen and rugby-league hall-of-famer, in the England Saxons set-up, and now wanted him as his right-hand man. I could see his logic. Andy was a strong personality and had an instinctive understanding of the realities of Test rugby, whatever the code. Rather like Martin Johnson, he was a root-and-branch realist who knew what it took to win games at international level. He had also been fast-tracked as a coach by Kevin Bowring, who was in no doubt as to his potential. These developments inevitably led to the 'exit stage left' treatment for John Wells and Mike Ford, who had been in place for two World Cups and three coaching cycles. It was my job to tell them they were no longer a part of England's plans, and then I headed to Saracens for a conversation with the chief executive Edward Griffiths over Andy's availability. Edward was extremely accommodating – he knew Andy was open to Stuart's overtures – and was happy to release his man for the duration of the Six Nations. Discussions would be far more vexed when we attempted to bring Andy to Twickenham on a full-time basis, but for the time being, we had what we wanted.

Instead of flying out in search of some winter rays, Stuart chose Leeds as the base for his pre-Six Nations camp. It happened to be on his doorstep and, if this was a major factor in his choice, he was far from the first England coach to make life easy for himself geographically: Brian Ashton had gone out of his way to base the team in Bath, partly because he lived just outside . . . Bath. But, in Stuart's case, there was more to his decision than mere convenience. It was another ingredient in the 'reconnection' mix. He wanted his team to be close to the rugby public, not distant from it. He wanted to hold open training sessions, fill some autograph books, give the supporters a stake in this new England team. And he was proved right. When we travelled to Scotland for the Six Nations opener with eight uncapped players, few in the media – or outside the media, come to that – gave us much of a chance. Yet we won, and then won again in Italy. Neither victory could be described as comfortable: had it not been for Charlie Hodgson and his penchant for charging down opposition clearance kicks, we might not have survived in either Edinburgh or Rome. But by reaching the mid-point of the tournament unbeaten after two proud and passionately determined performances, we'd done enough to convince Twickenham Man and Woman that this was an England team fit to take the field. Fit in mind and spirit, as well as body.

We could have won the third game against Wales and were unlucky not to do so, but such was the public reaction to England's performance that day, we emerged ahead of the game even though we'd lost it. And then we won in France, with the Leicester flanker Tom Croft performing uncannily accurate impersonations of an Olympic sprint medallist with his gallops in open field, before beating Ireland in convincing fashion at home to bring to a close a memorable few weeks. These last two games had been framed in the press as a Lancaster job application; while that was overly simplistic, there was no doubting the momentum that Stuart generated over the course of those eight days in March. The supporters felt they were setting out on an exciting journey towards the 2015 World Cup and, if I'm honest, so did I. Stuart had gauged the public mood perfectly and had done everything we could have asked of him.

We would have needed a very good reason not to upgrade his interim position to a permanent one, and while both Nick Mallett and Jake White continued to express an interest in the job, events had conspired against them. We interviewed Nick, even though he had used his platform as a pundit on South

African television to effectively concede defeat, but the course was set. Could we have guaranteed that a change of direction away from Stuart would lead us to the Promised Land? Not at that point. Not after finishing a very strong second in a tournament that could not have been staged at a more difficult time for us.

We went through the proper process by setting up an appointment panel. I was on it, as was Ian Ritchie, now fully on board as the RFU's chief executive. The other members were Ian McGeechan, who sometimes seemed to be a full-time *éminence grise* but was in fact doing a day job as director of rugby at Bath; Conor O'Shea, now at Harlequins and well known to the union; and Richard Hill, the World Cup-winning England flanker who was at that time working on player development at Saracens.

It was a modern set-up, designed to reflect the burgeoning partnership between club and country, and there were no great stresses or strains over the course of our deliberations. Stuart had auditioned so well, he could have won a scholarship to RADA: even the most enthusiastic Mallett backers could appreciate the strength of his position. He'd been put on trial and come through it. He'd removed players he felt had betrayed the shirt, made shrewd selection moves both on and off the field, and struck a good balance between immediate performance and building for the future. Ian Ritchie, who had made the jump from running a two-week tennis tournament at Wimbledon to presiding over an organisation that had been among the most dysfunctional in world sport for longer than anyone cared to remember, must have been thinking: 'This is a good gig after all. What was all the fuss about?'

Of course, England rugby would not be England rugby if everything stayed on track. The period between the victory over Ireland and the flight to South Africa for our three-Test series with the Springboks was marked by Andy Farrell's return to Saracens (as agreed with the club) and the saga surrounding Wayne Smith, whose name had routinely been included among the runners and riders for the head coach role before Christmas, even though he was never interested in heading things up at Twickenham. The business with Andy was a little strange initially, in that it was difficult to work out exactly what he wanted to do. Did he really feel he was not yet equipped to commit himself to England in a full-time capacity, as he suggested in his public statements, or was he simply showing loyalty to Saracens by returning to the club scene? As a man of honour he would certainly have felt conflicted – Sarries had given him his opportunity, after all – but, as far as I could see, he had revelled in the many and varied challenges of Test preparation. Stuart felt the same way and spoke to him throughout the episode. It was not long before both of us knew that given a straight choice, Andy would stick with England.

It was an awkward situation, however. Ed Griffiths was not playing quite so nicely now; there was a fair bit of angst in the air and, try as I might, there was no way of resolving the issue before the squad left on their Johannesburg-bound plane. In the end, though, Saracens knew they could not hold their man. It came down to money, as it usually does in negotiations with clubs, and by early summer Andy was sticking his thumbprint on an RFU contract.

With his closest colleague temporarily off-limits, Stuart felt he needed coaching support in South Africa. We turned to Mike Catt, who had been a part of the World Cup-winning squad in 2003, played in the losing final four years later, and had spent the latter years of his career at London Irish developing a new set of training-ground skills. He was keen and he was available, so we signed him up, initially on a short-term deal. Unfortunately for us, it transpired at around the same time that Wayne Smith was neither keen nor available, although it was a close-run thing. His decision to stay in New Zealand was one of the watershed moments in our build-up to the home World Cup, and while it is stretching a point to suggest that he would have helped us win the title in 2015, I believe deep down that we would have made a far better fist of the tournament had we succeeded in luring him to Twickenham. Not to put too fine a point on it, he would have made an enormous difference.

There had been a lot of talk, much of it overblown, about Wayne coming to England in the immediate aftermath of the 2011 shambles (not that it had been a shambles for him, given his immense contribution to the All Black victory in that tournament). But once the Six Nations had ended and Andy had made his excuses and left, Stuart was keen for us to keep the communication lines open. He felt the coaching team would benefit from a little more 'grey hair', and I could see where he was coming from – not because Stuart had very little hair of any description, but because Wayne was a coach of such stature. He was rugby gravitas in human form yet, being virtually free of ego, he was not the sort to be the 'big I am' in a coaching team.

The germ of the idea of bringing him to Twickenham lay in the now infamous review of our 2011 World Cup campaign. We had a range of discussions around the central topic of how to ensure that no England team would suffer such public humiliation ever again and, at first, answers there came none. But in the course of those conversations, we obtained a copy of the New Zealanders' report into their own World Cup failure in 2007. Contained in that document was a comprehensive account of the actions they had taken following the 2003 tournament and its immediate aftermath, during which they had some alarming disciplinary issues of their own. It was at that point that Wayne dropped a note to Graham Henry, the head coach, saying 'fix this thing': a clear hint that unless the All Blacks cleaned up their act, he would have nothing more to do with them.

This fascinated us. Apart from his huge expertise as a rugby strategist, he had direct experience of addressing a precipitous decline in standards and restoring a sense of decency and integrity to a leading national team. In the years following, those cultural problems were successfully addressed: the All Blacks' failure in 2007 was down to a leadership problem, not a behavioural one. By the time they reached 2011, everything was coming together and they duly won the title.

During that tournament, where he and Steve Hansen had been key figures in Henry's supercharged think-tank of all the talents, Wayne had made it clear that his long association with the All Blacks was coming to an end; that he wanted out of the silver-ferned rat race; that he intended to spend some time reacquainting himself with his family; that he would consider his options in a leisurely manner.

When we started speaking to him in earnest, he had taken on a Super Rugby role with the Hamilton-based Waikato Chiefs. The fact that his elderly parents were living in the area played no small part in his decision, but it was also true to say that his new contract contained an exit clause. If he so wanted, he could be available in time for the new northern hemisphere season. I spoke to Wayne a good deal and he was straight with me from the start, telling me that he would find it a really tough decision to leave New Zealand for England: partly because of his family circumstances and partly because the All Blacks were very close to his heart and he was not quite sure how much he wanted to coach against them with a team capable of threatening their domination of the world game. But equally, I knew he and his wife liked living in England – they had certainly enjoyed their stay in the Midlands during his time at Northampton – and that he had a good deal of respect for what we were trying to achieve and how we were going about it. Increasingly, we saw him as our missing ingredient: his signature was the big prize and I chased it as hard as I could. As did Stuart, who made a return visit to South Africa to talk things through with Wayne while the Chiefs were on a Super Rugby trip to the republic.

At one point, I genuinely thought I'd landed him: we were happy to give him the rest of the year off and receive him with open arms at the start of 2013, and even though the All Black hierarchy was placing him under huge pressure to stay put, we actually agreed financial terms. But in the end, I couldn't quite close the deal. 'I just can't do it,' he told me. And of course, he was back in the New Zealand coaching team for the 2015 World Cup in England. And who won the thing? Well, it wasn't us.

What might we have achieved had Wayne decided differently and brought all his experience and perspective to bear on the England environment? It's hard to say with any certainty, but he would surely have saved Stuart and the rest of the coaches from themselves during the run-in to the big event, when the good habits and sound management of the previous three years appeared to evaporate. The heat and intensity of a World Cup on English soil undoubtedly had its effect on Stuart, who flew directly in the face of his own good judgement at important moments and ended up paying a heavy price. With Wayne there to support him, things might have turned out differently. Wayne might even have talked Stuart out of shouldering additional responsibilities just when he should have been narrowing his focus and directing it solely on the one thing that mattered: the national team and its performance on the global stage.

Stuart might have come into the job to the sound of quiet applause rather than blazing fanfares, but with his great powers of diligence and huge capacity for hard work, he was not shy of extending his influence across ever larger swathes of the performance department. One of the legacies of the 2011 World Cup review was a recommendation for yet another review, this time of the entire professional rugby operation being run out of Twickenham. Martyn Thomas had come up with the notion during his time as acting chief executive and, for good or ill, it survived his fall from grace. By the middle of 2012, by which time we had rediscovered a good deal of our self-respect by finishing second in the Six Nations and pushing the Springboks in South Africa, the idea was still hanging around, so it fell to Ian McGeechan (definitely a 'usual suspect' by this stage) and Peter Keen, whose senior role at UK Sport put him close to the heart of strategy for the London Olympics, to go through things with a fine-tooth comb. Their findings put us back in 'performance director' territory and it was decided that Stuart should take on a significantly broader role while continuing as head coach of the England team.

It was a bad move, both for Stuart and for the union. I had no particular problem with the undertaking of the review, although I remember wondering how many of these damned things we would have to go through: my remit was extremely wide-ranging and with negotiations about to begin on a new long-term deal between the RFU and the Premiership clubs, I would be busy enough. But I had serious reservations about Stuart spreading himself so thinly just when he needed to be concentrating 24 hours a day on preparing the national side for the biggest tournament of a rugby lifetime. No head coach had ever been given such complete control and, sure enough, it backfired. Instead of concentrating on a single, overwhelmingly important project, on agenda item No. 1, Stuart became tangled up in all sorts of extraneous matters. This dipping in and out of subject areas of minimal relevance to the primary task in hand was a distraction that could and should have been avoided. Just when we needed clarity, we had distortion; just when we needed a clear road, we had unnecessary clutter.

Did the famous victory over the All Blacks in early December 2012 affect the thinking and drive the policy at Twickenham? It must have played a part. If the heart-warming England recovery during the Six Nations had put the rugby public firmly in Stuart's camp, many supporters were prepared to canonise him after a three-try performance against the reigning world champions that will live long in the memory. The head coach could have run for President that Saturday night and been voted in by a landslide. A little over a year on from the carnage of Auckland and its depressing aftermath, there were legitimate reasons to feel good about England rugby once again. If senior RFU figures allowed themselves to get a little carried away by making him master of all he surveyed – well, that's human nature.

As it turned out, the on-field progress continued for a while. There was an 80 per cent winning return in 2013 and, if the percentage return slipped rather dramatically in 2014, there were four meetings with the All Blacks in that calendar year, three of them in New Zealand – territory so inhospitable that we had won there only twice in our entire history. England finished on the losing side in each of those match-ups but, leaving aside a heavy defeat in Hamilton in the last of the summer Tests, a total of just nine points separated the teams in Auckland, Dunedin and London. On balance, things could have been a lot worse, especially as we knew that with the grace of God and a following wind, we would not meet Richie McCaw and Daniel Carter again until the final of the World Cup.

Even the blips elsewhere could be rationalised away. There was a horrible defeat in Cardiff on Grand Slam day in 2013, but when England crossed the Severn Bridge again at the start of the 2015 Six Nations, we delivered a high-quality performance to win 21-16 – a margin that did not even begin to reflect our superiority. We lost by two points in Paris in 2014, but scored 50-odd points against the French at Twickenham a year later. We lost narrowly to the Springboks a year out from the World Cup, but beat Australia by more than a score a couple of weeks later, which seemed far more relevant given that England would be sharing a World Cup pool with the Wallabies. Victory in that group would give us a clear route to the final: Scotland in the last eight, Ireland or Argentina in the last four. Neither game would be easy, but let's face it: who could have asked for anything better?

If there was a problem team-wise, a selectorial Achilles heel, it was to be found in midfield. This had been a running sore since the break-up of the World Cup-winning combination of a dozen years previously: Jonny Wilkinson, Will Greenwood and the two Mikes, Tindall and Catt, who had contested the outside-centre berth during the tournament in Australia. Back in my day under the stewardship of Geoff Cooke, there was something close to complete certainty over numbers 10 to 13: I was the outside-half, Will Carling and Jeremy Guscott were the other men in the equation. It was very different now. Try as he might, Stuart could not settle on an optimum configuration. Which is where Sam Burgess came in, and where things went horribly wrong.

I would not even begin to pin the blame for our embarrassing World Cup misfire on a single player, but the kerfuffle around the introduction of Burgess was undeniably the tipping point. To this day, I simply do not understand the thinking behind the fast-tracking of a player from international rugby league to international rugby union when so many of the things that had made him wildly successful in the 13-man game were of questionable relevance in the

15-man version. It was an almighty risk to select him in a World Cup squad on such extremely limited and highly questionable evidence and it proved to be an almighty blunder. Stuart and Andy Farrell have always defended their position on this, but as far as I'm concerned they can say what they like: Burgess was a rogue ingredient in the mix, both before the tournament and during it, and his inclusion had a negative effect.

Why did Stuart do it? He alone knows the truth of the matter. But all head coaches are control freaks in their own ways, especially around the matches and tournaments they know will define them, and Stuart became pretty dictatorial in the way he ran the show in 2015. The Burgess business revealed him at his most obsessive: he was clearly not happy with his options at No. 12 and had made up his mind that Sam offered him the nearest thing to a way out, despite the reservations of those who had not seen anything from him at club level with Bath to suggest that he was even remotely up to speed with the realities of midfield play at Test level. Stuart's *idée fixe*, his infatuation, led to all manner of conspiracy theories: some people were convinced that the RFU was paying a chunk of Burgess's very handsome salary (we weren't, even though the Bath owner, Bruce Craig, came to us in search of a contribution); others said there was some kind of pressure to pick him from our commercial partners (as far as I know, there wasn't). To my mind, it was a simple error of judgement: simple, and enormously costly. Stuart had made such a big thing about the importance of fighting for the honour to wear the shirt, yet in the frenzy of an impending World Cup he allowed something to take root in his mind that directly contradicted the very principles on which he had rebuilt the culture of the team. It was massively unsettling for a bunch of willing individuals who had been through a fair bit since coming together and had become very tight-knit as a playing group.

All of this brings me back to Wayne Smith. I don't think for a moment that the Burgess thing would have been allowed to take on a life of its own in the way it did had Wayne been on hand to keep things in perspective. I also think our physical preparation for the tournament would have been very different if he had been around. After the last round of Six Nations matches, when we almost pinched the title in what amounted to a cricket-style run chase against the French and left the Twickenham crowd in a state of feverish excitement the like of which I'd never previously witnessed, England had travelled to the United States for some high-altitude training in Denver, Colorado. Too many players went on that trip: instead of being a highly detailed, narrowly focused, tournament-specific camp, it bore all the hallmarks of an extended trial geared towards final selection. Those involved did an awful lot of running, on the presumption that the World Cup matches would feature unusually high levels of ball-in-play time (which they didn't). Even when the wider squad gathered again for the pre-tournament warm-up games with

Ireland and France, there was still uncertainty over selection. Why did Danny Cipriani play against the French in Paris when his chances of surviving the final cut were seen as somewhere close to zero? Why was there a competitive trial game at the team base in Surrey the day before the final announcement? Why did some players struggle to sleep that night for fear of being axed at the last minute? This was the polar opposite to the All Black way of doing things. Where was Wayne when we needed him?

A couple of weeks before England's opening tournament match against a dangerous Fiji side full of outstanding rugby athletes, I had another of my brushes with the press. Actually, it was more than a brush. I was vilified. The flare-up centred on my interview with the *Daily Telegraph* sportswriter Paul Hayward, who asked me if I felt the England team was peaking for the forthcoming competition in the way Clive Woodward's side had peaked in 2003. I said I doubted it. 'I suspect this team will get better over the next two or three years,' I continued. 'I don't think there's any question of that, because the age profile and the experience profile is going to grow.'

Well, pardon me for being honest. I said what I thought and, if truth be told, I think my words were a pretty accurate reflection of what the head coach was thinking too, although Stuart was not the one going on the record. I went on to say that this didn't mean we couldn't be good enough over the next few weeks, but when the other papers picked up on the comment, some of them were slightly selective in their accounts. According to the headline in the *Daily Mail*, I had made an 'astonishing gaffe'. To my mind, it was neither 'astonishing' nor a 'gaffe'; rather, it was a straightforward assessment of our position and a long way short of rocket science. Compared with the avalanche of 'Go Now' bile that had swept over me at the back end of 2011, this was not much to write home about. It did, however, serve to remind me that I still had my foes among the chattering classes. If they weren't all out to get me, it didn't mean that none of them were.

Having lived through the high emotions of the World Cups in 2007 and 2011, there was nothing in the 2015 version that had a similar impact on me, largely because I was nowhere near as connected to events as they unfolded. My overriding feeling as we paid the price of our muddle-headed defeat by Wales and our no-show against the Wallabies was one of real sorrow. I felt for the people, coaches and players, who were crushed by the outcome, and I thought it a crying shame that English rugby had let such a wonderful opportunity slip through its fingers. It seemed to me that we froze: that the pressure was too great for those at the sharp end to bear. How else to explain the nonsense at the end of the Wales game, when we turned down a penalty shot that would have drawn the game in favour of a line-out call that was risky at best and spellbindingly dumb at worst? On this subject, I'm about as old school as it gets: in Test rugby, you take your points as and when they arise. Always. End of. Finish. The statistics around kicking to the corner rather than kicking at the sticks can be read in only one way, and they don't favour a hit-or-miss throw to the line. Wide-angled as the position was, there was barely a person in the ground that night who would not have backed Owen Farrell to nail the three points on offer. Instead, we were treated to a slow-motion car crash – a scrambling of the decision-making process, an unravelling of everything we had worked for since that Six Nations camp in Leeds in the cold early weeks of 2012. At that moment, our chances were dust. The valedictory capitulation against the Australians a week later was entirely predictable.

How did the RFU react this time? With another review, of course. Thankfully, this one had nothing to do with me – not just because the business of the national team was now outside my remit, but also because I was not called 'Ian'. The panel consisted of Ian Ritchie as CEO, Ian McGeechan, Ian Metcalfe and Ian Watmore, a successful management consultant and former senior civil servant who had served on the England Rugby 2015 board. Ben Kay, the World Cup-winning lock from 2003 who had also played in the 2007 final, was drafted as the fifth member, presumably for reasons of variety. It was a delicate undertaking for all concerned – some players, mindful of the 2011 episode, initially expressed a reluctance to provide feedback – and great care was taken to avoid a repeat of the leak that had left such a stain on the sport. Things moved quickly, even so. Within a month of the review being launched, Stuart was on his way out of Twickenham.

My view of Stuart Lancaster now? He was undone by his workload – a self-inflicted wound in many ways. I wouldn't expect him to agree with the analysis, but I believe he took his eye off the ball. The head coach of a national team aiming towards a World Cup is a big job in whatever context you care to name. You have to devote 100 per cent of yourself to getting selection right because that's the number-one skill; you have to ensure that you and your coaching team are preparing the side in the best possible manner to win the next game. Yes, there are times when you should also be looking at the bigger picture, and Stuart was right to do so at the start of his tenure, when expectations were low and he had room to manoeuvre. But by 2015, things were different. He did a fine job in rebuilding the team but, for me, he had to sharpen his focus 18 months before the big event; he had to move from culture-building in the broad sense to creating a winning culture in the narrow sense. Everything in his working life should have been about one single thing, and it wasn't.

Yet whenever I reflect on that World Cup, I find myself smiling through the lingering disappointment. English rugby had produced a stunning tournament that heightened my senses from start to finish. I had watched France play Canada at Milton Keynes in front of 30,000 on a Thursday night – and a wet Thursday night at that. I'd seen Australia's second string in a turkey shoot against Uruguay at midday on a Sunday at a packed Villa Park, and revelled in every last second of it. I'd been back to Newcastle for New Zealand v.

Tonga; to Cardiff in midweek to see the Wallabies take on Fiji; to a sold-out Wembley to cast an eye over the Argentines' brave assault on the All Blacks; to the Olympic Stadium for Ireland v. Italy, where a dire contest was enriched by the quality of the match-day experience. I'll be long dead before I forget the colour and din of Kingsholm in Gloucester on Argentina v. Georgia day – one of the most phenomenal rugby occasions of my life. There was never a hint of despair in the air, even when I found myself watching the Wallabies at Twickenham on five successive weekends. This was feast, not famine.

And of all the assaults on my emotions, the one that lives with me to this day is the chilling eeriness around south-west London after the England v. Wales game. I found myself walking out of the stadium in the direction of Twickenham Station at 11 that night, in search of a taxi ride back to my home in Kingston upon Thames. There were hundreds, perhaps thousands of Welsh supporters, men and women and red-shirted kids, walking the same route, but there was no great sense of joyousness about them. They were in shock, just as I was. They'd mugged us and they couldn't quite believe it, any more than I could. I will remember their silence for ever and a day.

12
OVER AND OUT

THERE WERE PLENTY of bitter jokes flying around after the national team's tail-between-legs retreat from their own World Cup, one of which went something like this: 'All things must pass . . . except the England midfield.' If there was an 'ouch' factor attached to each and every one of the smart-arsed one-liners, did any of us employed in the front line at Twickenham have any just cause for complaint? Many true words are spoken in jest, as most of us have found to our cost at one time or another, and while my direct influence on recent events had been practically zero, it was important for everyone involved with the RFU to present a united front. Rugby union is built on several articles of faith, the most important of which states that we win together and we lose together. For this reason, among others, I was mightily relieved that the 2015 fallout was infinitely less toxic than the 2011 version.

At the same time, I was asking myself some hard questions about the future. This self-examination had nothing to do with what had happened over the course of a blighted autumn: my unease was not the result of falling out of love with, or losing interest in, the England team. If I'd come to understand anything about myself since crossing over from club to country, it was that I cared about our fortunes at Test level even more than I'd previously realised. But the most pressing part of my job, the delivery of a new long-term agreement with Premiership Rugby, would be complete within a few months, by which time I'd have spent almost a decade in the same place, if not quite in the same role. 'What then?' I thought. 'What's next? Is there a next?'

Without wishing to sound overly dramatic or sentimental, I see myself as a doer, not a dreamer – my heart was telling me I was ready to go. To say I felt I'd done my time would not be accurate: there had been any number of difficult moments over the previous nine years or so, but I hadn't been serving a prison sentence. It was more a case of thinking: 'I've seen more in this job, good and bad, than I could possibly have imagined. Is there anything more to see?' Once I'd decided that the answer was 'probably not', the rest fell into place.

When Ian Ritchie arrived from Wimbledon as our new chief executive in 2012, a few weeks after the RFU had hit its lowest point since the great split with the north over broken-time payments and the creation of rugby league in 1895 (I sincerely believed it was that bad at the time and still believe it now), I had an open and honest conversation with him.

'If you want me to stay on and continue what I'm doing,' I said, 'that's fine. If you don't, that's fine too. If you think I should go now, I'll go.' Ian was completely upfront with me, as I was with him. He was keen that I should stay on and help see the union through to the home World Cup. After that, he said, I'd be free to do as I wished. I think I knew then that 2016 would be it for me: that with a second eight-year agreement with the clubs in the bag, I could walk away satisfied that I'd been instrumental in negotiating 16 years' worth of stability for the elite professional game in England. I really didn't want to go through another Martyn Thomas-type episode, so I promised myself that I'd be the one in control of my own departure. And that is exactly how it panned out. Knowing that the club deal would be finalised in June and that it would run right the way through to 2024, I announced in April that I would be heading for the exit at the end of July.

After the tough talks with Tom Walkinshaw in 2006 and 2007, the discussions with his successor as the Premiership's negotiator-in-chief, Bruce Craig of Bath, were nowhere near as fraught. This is not to suggest that Bruce was a pushover. Far from it. From the moment in 2010 that he bought the West Country club from Andrew Brownsword (a very different character, to the point of being a polar opposite), he had been among the most prominent and outspoken of the top flight's owner-investor-chairmen. Unlike Brownsword, he had a lifelong interest in the sport and was extremely ambitious in pursuing the Premiership agenda. That's unlikely to change: now that I'm looking at the sport from the outside, I'll be very interested to see how hard he pushes some of his more radical ideas for the development of the club game. But over the period of our time together around the table, we dealt largely in tweaks, not in upheavals. The club-versus-country politics were nowhere near as raw as they had been in the late Noughties – a sure sign that the original agreement had struck the right chord and served its purpose – and, as a result, the progress we made was steady, if not entirely painless. The big issue of control, of 'ownership' of the game, was not on the agenda in the way it had been, primarily because there was now a recognition that no such thing could be achieved. So it quickly came down to money, with Bruce effectively saying: 'What do you want from us and how much are you willing to pay for it?' In return for a continuation of our quasi-central contract arrangement without central contracts, we coughed up more cash. Double the amount of cash, actually: a figure broadly in line with the upward curve in RFU revenues. There were some slight variations to the original deal, but they were to do with shifting the emphasis towards the England Qualified Player scheme rather than anything structural. It was a sign of the times. Post-Brexit and the loosening of our ties to European legislation, cash incentives for picking home-grown talent rather than foreign imports will no longer be the complex and sensitive subject it once was.

One of the principal beneficiaries of the continuing peace and stability should be Eddie Jones, the first foreign boss of the England team. Several conclusions were reached in the aftermath of the World Cup, the first of them being that Stuart

Lancaster could not continue. The second was that we should not continue along the same one-way 'England for the English' road we had been following for the thick end of half a century. It was way back in 1969 that Don White, born and bred in rural Northamptonshire, became the first head coach of the national side, and while some of those who succeeded him had the full range of credentials and were therefore natural appointments, there had been nothing quite so straightforward in recent times. I had turned to Brian Ashton in an hour of need; both Martin Johnson and Stuart had to a large extent been shots in the dark. We were done with experimenting now. We needed the right man – the very best available – and if that meant making that man and his existing employers offers they couldn't refuse, so be it. Short of leaving a horse's head in someone's bed, we were prepared to do whatever it took.

Who might the right man be? Someone with a proven track record at international level: someone with a gift for finding the pressure points of his players and pushing the right buttons; for giving clear messages in terms of tactics and strategy; for selecting the best personnel in their optimum positions. There was no swamp-draining to be done: to his credit, Stuart had accomplished a great deal in this respect. This was about performance, pure and simple. We were looking for a coach who backed himself to deliver on the field, time after time. It was on this basis that those of us charged with identifying the prime candidate cast an eye over the possibilities. Who were the Big Men?

There was Steve Hansen, of course, but he had just won the world title with New Zealand and had a British and Irish Lions tour ahead of him. Wayne Smith? Been there and failed. Could we buy Warren Gatland out of Wales? Perilously difficult, diplomatically as much as financially. Joe Schmidt out of Ireland? Equally tricky. When we really thought it through, we were counting the contenders on one finger. It had to be Eddie. It was almost a no-brainer. He'd taken an undercooked Wallaby side to a World Cup final in 2003, helped the Springboks to the crown in 2007 and spent long enough in the Premiership with Saracens to understand how things worked in England. He had also achieved scarcely believable things with Japan only a few weeks previously. Had they lost all their World Cup games by 30 points or more, his star might have been in the descendant: instead, they had won three of their four matches – one of them against the mighty South Africa – and gone within a hair's breadth of reaching the knock-out stage.

Eddie's star was radiating light and heat like never before, so Ian took an early flight to Cape Town, where our number one target was just starting a tour of Super Rugby duty with the Stormers, and talked him back across the Equator.

It was clear that the RFU was breaking new ground with this appointment: we'd had southern hemisphere coaches on the England staff – John Mitchell, Brian Smith – but we'd never brought one in at the top. Yet there was no English alternative available, as far as I could see. After what had happened, was it really open to us to give the job to a Dean Richards or a

Jim Mallinder or a Steve Diamond, none of whom had run the show at any serious level of the international game? For all his eye-catching achievements at Exeter in recent seasons, could we really turn to Rob Baxter? As I've already explained, England is not New Zealand. We're not in a position to encourage a Baxter or a Diamond to get out there and see the world, to experience a different rugby culture, to book themselves into a finishing school in Auckland or Wellington or somewhere in the wilds of Otago. Our game is not set up that way and probably never will be. Even if we found a way of facilitating moves abroad, would our home-bred coaches want to go anyway? The salaries are good in the Premiership; very good in many cases. In some important ways, we have more in common with football than we do with All Blacks rugby.

High on my list of concerns about the direction of the sport I love is northern-hemisphere rugby's tightening embrace of the football model. I am not suggesting for a minute that the union game in this part of the world should follow the example of my other great sporting love. Leaving aside the extraordinary Twenty20 phenomenon of the Indian Premier League and one or two copycat productions around the globe, cricket is driven entirely by the finances generated in the international arena. The elite Test performers hold central contracts and are, to all intents and purposes, full-time England players. It is a top-down arrangement, pure and simple: the county game survives in its current form because the international game makes money. Could this work in rugby union?

Absolutely not. Not now. It might have done, had the RFU been even half awake in 1995 and shaped the newly professionalised sport to its own design, but that is like saying we'd have beaten the All Blacks that summer if Jonah Lomu had been six stones lighter and only half as fast. The front-rank clubs in England have grown too strong to be wished away by the nostalgia-soaked committee men from the shires, just as they have in France.

The question is not whether the big teams on either side of the Channel will eventually run out of cash and throw themselves on the mercy of their respective governing bodies, but at what point they decide that they have sufficient money of their own to disengage with those bodies in the way Premier League clubs cut the cord with the football establishment back in the early 1990s. At the moment, they are between a rock and a hard place: they want to stand on their own two feet, but the arithmetic does not support a unilateral declaration of independence. However, they are closer now than they were when I was at Newcastle, and they will be closer still in a few years' time, always assuming they do not blow all their resources on players' wages. This story has the potential to run and run.

I've been on both sides of the fence and I know this much: it is only a matter of time before the next big battle between the independently run clubs on the one hand and the governing bodies on the other. And before anyone down south or in the Celtic lands makes the mistake of dismissing events in

England and France as local squabbles, they should understand that we are talking here about the two most lucrative markets in the world game: indeed, the only markets that matter a bean in the wider commercial sense. Without political vision and diplomatic sure-footedness on the part of the various stakeholders – administrators, owners, coaches, players – the non-alignment of interests and inherent conflicts at the heart of the two biggest rugby nations on the planet could easily bring the entire sport to its knees. It is plain to see that the most powerful European clubs are attempting to minimise their overlaps with the international programme, hence the muscle-flexing over the most recent attempts to bring north and south together in a 'global' season. Can we blame the Bruce Craigs and Nigel Wrays of this world if they're running out of steam with the idea that Bath and Saracens should meet on serious Premiership business on a Test weekend, without their major box-office attractions? Of course not. These people have funnelled vast amounts of money into rugby union: it must have cost Nigel well over £25 million if it's cost him a penny and he still spends an unhealthy chunk of the season watching his team play important fixtures while stripped bare of their top-of-the-bill attractions. Yet it is equally true to say that the calendar year consists of 52 weeks, not 72 weeks. The most you can squeeze into a pint pot is a pint. Until that changes, which it won't, there will always be more rugby to be played than there is time to play it in.

Unless, of course, someone makes a sacrifice by giving

ground, which is not something we have seen too often in the last 20 years. I do not believe we are watching a fight for total power any longer: in English and French terms, that battle was lost by both sides a good while ago. Where rugby finds itself now is in the early stages of a conflict over the nature of the compromise – a scrap for viability. The main theatre of action may be Europe, but there is not a corner of the union landscape that can claim to be unaffected. Not even New Zealand, who may be delivering with staggering consistency on the field but cannot claim to be quite as successful off it. If they are staying afloat financially, it is because they make themselves available for hire at the right price. (I very much doubt that their 2016 game against Ireland at Soldier Field in Chicago, of all places on God's earth, was a charitable event.) Their problem is that 'chequebook rugby' cuts both ways. What happens when Wasps or Bath or Toulon or Montpellier or Racing 92 start offering seven-figure annual salaries to the stellar All Blacks rather than the mere £500,000 or £750,000 they are shelling out these days? How deep will the attachment to the silver fern run then? The New Zealanders will have to find sufficient dollars to remain competitive as best they can, otherwise the entire foundation of their domestic game will be weakened. But, by doing so, they could easily go bust. It's the same in Australia and South Africa, and Ireland and Wales, come to that. Without compromise, the rugby union model worldwide will be under threat.

In one sense, professional rugby union has become a

well-established part of our sporting lives: the big Six Nations and World Cup games draw spectacular audiences on terrestrial television, and if elite club rugby in Europe has less of a presence in the free-to-air market, its footprint is immeasurably greater than it was when I sat down in that London restaurant with the money men from Tyneside in the late summer of 1995. Yet, in another sense, it is still in its infancy. If some of those who bought into the project right from the start imagined they would receive a speedy return on their investments, the majority of us who had first-hand experience of rugby governance and understood its structures knew that the idea of a quick fix was illusory. The journey from the Wild West to Shangri-La – from borderline anarchy to a fully productive, self-sustaining, grown-up sporting business – was always going to be fairly arduous. In my judgement, the game is now in a hybrid state: it doesn't quite know where it is, or what it is. The international game continues to hold the stronger hand, just about, but while governing bodies can still say to the clubs, 'We're having your best players for Test matches on these dates and we're throwing some money at you as compensation', they are doing it from a position that grows ever weaker. Where once they ruled through regulation, they now rule only through the chequebook. To my mind, that is a very significant shift.

It was no easy matter, reaching an accommodation with the Premiership negotiators on behalf of the RFU: the talks were long and drawn out, heavily detailed and frequently exasperating. But that's the nature of compromise. My concern is that people on both sides of the club–country divide are still prone to the odd knee-jerk reaction. For instance, how did the top brass at Twickenham arrive at the conclusion that a hastily arranged 2017 autumn Test against the All Blacks, in addition to the agreed fixture list, was good politics? It was not quite a mirror image of the 2006 affair, but there was still potential for trouble and the abrupt abandonment of the plan was a good move. How did the financiers of the two major Paris teams, Racing 92 and Stade Français, even begin to imagine that a sudden merger might be acceptable to the players and supporters who are the lifeblood of the game in the French capital? (That plan failed after a strike threat.) And when it comes to Super Rugby in the southern hemisphere, on what planet do the organisers live? At the back end of the amateur era, there was a Super 6: three teams from New Zealand, two from Australia and – how open-minded this seems now – Fiji. The return of a prodigal son in the shape of post-apartheid South Africa resulted in an expansion to Super 10: then there was a Super 12, a Super 14 and a Super 15, before the inclusion of Argentine and Japanese franchises gave us a Super 18, which featured a mind-bogglingly complicated regular season and a set of geographical problems that would have driven Marco Polo himself into the arms of a therapist. I admire Argentine rugby as much as the next man, but Buenos Aires is a hell of a long way to go for 80 minutes of rugby.

The southern hemisphere game likes to promote itself as a paragon of stability, a bulwark against the ruinous excesses of club rugby in Europe. The truth is very different. Do South African rugby folk love Super Rugby in the way they loved, and still love, the domestic Currie Cup? Do supporters in New Zealand really identify with the Hurricanes or the Chiefs in the way they identified with Wellington or Waikato?

To a large extent, professional rugby union is a victim of its own progress. The game I played in the closing decade of the amateur era was completely different from the one we watch and marvel at today. By way of emphasis, I might add that the pace and physicality of today's product at Premiership level is immeasurably greater now than it was when I called time on my career after that training-ground injury in September 1999, a point at which the weekly wage had already been a part of our lives for four years. Yet we continue to demand more of the players: more rugby in more months of the year, played at ever-higher velocity, at ever-greater risk to life and limb. The reason? No one wants to cut back on the number of matches, because matches mean money. I am not one of life's natural revolutionaries, even though I loved my time on the barricades at Newcastle, but there is a part of me that wonders if the situation can possibly be resolved without some form of industrial action from the workers who provide the entertainment.

To add fuel to the fire, there is no obvious way of depowering the game without making it unrecognisable: players will always run and tackle as fast and as hard as they can, and if increasingly sophisticated developments in sports science make them quicker and more powerful still, there is only one outcome. It is for this reason that football's solution to fixture planning – the staging of major matches in midweek as well as at weekends – is a non-starter. Yes, we routinely turned out on Tuesday and Wednesday nights when I first went into adult rugby, but the creation of the English leagues in the late 1980s put paid to that. Back in the day, half the fixture list was reassuringly gentle; now, there are serious athletes everywhere you look. In this respect, the dial turns only one way. So the search for an answer rests not with the style and substance of the rugby, but with the amount of rugby. Should there be fewer internationals? Probably, but who wants to take the inevitable financial hit? Should the Premiership be cut from 12 teams to ten? Ideally, yes. I have long believed that figure to be the optimum, in commercial terms as well as on the player welfare front. But how do you get there without imposing a franchise system? And if you go down that road, who will clean the blood off the carpet?

And yet, I find reasons for optimism. Why? Because if I've learned one thing on my 20-year journey through the professional game, it is that rugby union is incredibly resilient – a sport blessed with the most extraordinary ability to survive whatever may be chucked at it. What lies at the heart of this resilience? I keep coming back to the players. Whether it be a Premiership Grand Final or a European showpiece or a World Cup knock-out tie, the direct participants keep on finding ways of delivering something that captivates the paying public in the stands and the millions who watch from their own sofas. This is not to suggest for a second that the people who administer and finance the sport can be at all complacent: legally and medically, in player terms and in spectator terms, the challenge of keeping the union game healthy, vibrant and relevant will become more difficult for as long as it is possible to foresee, not easier. When you get right down to it, rugby's essence, its defining characteristic, is to be found in the single word 'contest': no other team game devised by man is based so squarely on the fight for possession in every facet of play. We do not live in an age of blood-and-guts heroism without consequences, however; we live in an age of regulation and litigation. The union game has its enemies, as well as its supporters, and it treads a fine line.

For all that, I believe in its future. My reasons are rooted in the past as I experienced it: particularly the recent past, where the essential values and qualities of the game – its soul and its spirit – held firm whatever the nature of the existential threat it encountered. I go back to those Twickenham weekends during the 2015 World Cup, when, despite the events surrounding the England team, I arrived home each night grateful for the mind-blowing sport I had just witnessed. If I ever doubted that rugby union could develop into a genuinely global attraction, the sight of so many full-house crowds, made up of so many nationalities, gave me the reassurance I needed. There was not a single whiff of trouble or animosity during that tournament: to walk back towards the centre of Twickenham after the defeat by Wales, to be recognised and serenaded by red-shirted visiting supporters who had drunk their fill and not to fear for a second that things might turn unpleasant – that's the magical side of the game. God forbid we ever lose it.

As rugby people, we recognise our self-destructive gene: we understand our capacity to make life difficult for ourselves by making the wrong decisions and fighting the wrong battles for the wrong reasons. But we also know that we are part of a game squarely based on goodwill and are therefore equally capable of finding solutions. Rugby's strength is in its nature. On the field, you're as competitive as it is humanly possible to be, and every now and again someone oversteps the mark. When that happens, you put it out of your mind: you get up and you go again, time after time. Is there any better way of living your life? No matter what the administrators and businessmen and coaches throw at it, rugby union will continue to be what it always was: it will forever come down to 30 people on a field – women as well as men these days, triumphantly – trying to dig their way out of a problem. Some of them will manage it and some of them won't, but the outcome is not really the point. What matters is the quest. For all the trials and tribulations of my life in rugby, I walked out of Twickenham knowing that in one overwhelmingly important respect, I was leaving the sport where I first found it. With its heart in the right place.

Postscript:

PUTTING A PRICE ON THE PRICELESS

The British and Irish Lions tour every four years: a simple statement of indisputable fact, the rugby equivalent of one plus one equalling two. So how is it that life with the Lions is so mind-bogglingly complicated? Blame politics and economics, those malign forces that disrupt the smooth running of virtually every top-level sport sooner or later and have bedevilled the game I love for just about as long as I can remember. The stresses and strains affecting the Lions are rugby's problems in microcosm. The global union calendar, control of the players, the divvying up of money, the welfare of the participants in both the short and long terms – all the major points of conflict in the sport in general are part of the Lions story in particular.

To use a familiar phrase from my Yorkshire childhood, I see trouble at t' mill ahead and it may not be long in coming. Before the Lions head off on their next jaunt across the Equator, to the land of the Springboks in 2021, we can expect some tough talks over the length of the trip – there is already a good deal of discussion about cutting the itinerary from ten matches to eight – and the amount of preparation time granted to the players ahead of the trip, not to mention the rest period they should expect after it. This alone has the capacity to change the face of Lions touring. The current ten-match programme just about allows the coaches to choose a relatively large squad and give everyone at least one full game to stake a claim for a Test spot. By chopping out two fixtures, you kiss goodbye to the old way of doing things. The Test team would have to be picked ahead of departure, and that would reduce the size of the party. Would an inexperienced outside back like Elliot Daly of England have made the cut for the recent visit to New Zealand in a slimmed-down party of 30? Probably not. As a result, there would have been no inter-continental penalty kick from the back end of beyond to keep the Lions afloat in the series decider in Auckland.

We can add to this the inevitable long discussion about financial compensation to the clubs providing the players: particularly the English clubs who, as independent entities free of direct union control, are in a position to drive a very hard bargain indeed. Throw in the growing clamour for the addition of Argentina to a roster currently restricted to New Zealand, Australia and South Africa and it is reasonable to suggest that the agenda is sufficiently demanding to test the most sophisticated of negotiators. As sophistication is some way down the list of virtues commonly associated with the people who run the sport – for some reason, the phrase 'tell me about it' springs to mind – things could become just a little lively.

But there will be a solution, reached in the time-honoured way: a good old scrap around the table and some spoutings-off in the press, followed by a financial settlement and tentative hugs all round. Why? Because there has to be. Different interest groups see the Lions in different ways, but only a fool would attempt to deny that the red-shirted brotherhood is already the biggest brand in the sport and shows every sign of growing bigger still. The Lions have two precious assets: a rich history rooted in the epic tours of the past – the great crusades of the amateur era that were measured in months rather than weeks and gave rise to so many legends and fables and tales of derring-do, culminating in the triumphs of the early 1970s – and a rarity value that sets them apart from all other teams in all other sports. Only with the Lions is there an air of pilgrimage among the supporters who travel with them, an idea of a popular front fighting the good fight in pursuit of an ideal. If the history goes back well over a century, the surge in interest is a product of the professional age and its impact in streamlining itineraries and making it possible for many thousands of rugby folk from all parts of the British Isles to follow tours from start to finish. That in turn has generated increased interest from the broadcasters, who now provide wall-to-wall coverage from day one, and as a consequence of their exposure, more and more sponsors and advertisers are clamouring to be a part of it.

The owners and chief executives of the Premiership clubs may be less than ecstatic at the demands placed on their most highly prized players in a Lions year – demands that are undeniably extreme – but they know a successful business model when they see one and understand that there is plenty in it for them if they play their cards sensibly. To my mind, it would be catastrophically short-sighted of them not to work with the sport's custodians at national and international levels to ease the pressure on the Lions and guarantee them the future the game at large patently wants to see.

For all the grumbling and sabre-rattling from the Premiership fraternity – senior figures at the new English champions, Exeter, and Leicester, still the biggest club in the country, were especially vocal on the subject of Lions reform as the latest vintage headed off to New Zealand for the 2017 challenge – most, if not all, of the powerbrokers in our domestic game recognise the strength of the brand. And for those out there who did not quite get it before the events that unfolded in the Test matches in Auckland and Wellington, they must sure as hell get it now. Here was another remarkable episode, bordering on the extraordinary: a stand-alone chapter in the history book, a story worthy of its place in Lions lore. Warren Gatland – only the second man after Sir Ian McGeechan to retain the head coaching job for consecutive tours – did not quite emulate the 1971 side in winning a series against the All Blacks in New Zealand, but he achieved the next best thing by squaring it. Many will argue that in doing so, he earned himself the right to be talked of in the same paragraph, if not in the same sentence, as Geech and (heresy of heresies) Carwyn James as a Lions coach for all the ages. I'm not entirely sure I'd place him in their company because the way I saw it, mistakes were made that cost a very good touring side the prize they craved. But only the grouchiest of curmudgeons would deny him a considerable degree of credit.

So what was it that we witnessed over the months of June and July? First and foremost, we saw the effects of pressure. If we look at the thing dispassionately, even the most partisan of Lions supporters – and I include myself among them, for having worn the Test shirt against the Wallabies in 1989 and against the All Blacks four years later, my feelings for the red shirt run very deep – we have to accept that the New Zealanders have only themselves to blame for not fulfilling pre-tour expectations and coming out comfortably on top. There were an awful lot of mistakes on both sides of the ledger, from individual drop-offs in skill levels to fundamental system errors: by the end, the teams reminded me of two drunks trying to hold each other up, such was the extent of the stumbling and slippage. But it was particularly surprising to see the home side, double world champions and barely beatable for an alarmingly long period of time, fall victim to outbreaks of butterfingered fumbling, wayward kicking and muddle-headed option-taking. Furthermore, they grew worse as the series progressed, regularly committing mortal rugby sins and then compounding them with others. Generally speaking, these series are meant to work the other way around.

In part, the All Blacks could claim to be victims of misfortune, some of it of their own making but most of it squarely in the sod's law tradition. The fact that they played two-thirds of the second Test in Wellington at a numerical disadvantage was down to the rank indiscipline of Sonny Bill Williams, their feted centre: it was his decision, and his alone, to smash the English wing Anthony Watson into the middle of the following week with a shoulder to the head. He was sent off and rightly so, but while there was no question as to his guilt – case open, case shut, over and done with – the French referee Jérôme Garcès might easily have taken refuge behind a yellow card and let the post-match disciplinary process take its course. All Black dismissals are rarer than hens' teeth: only two previous examples in the annals of international rugby, the most recent of them in 1967. Never before had it happened to a New Zealand player in New Zealand. Whichever way you cut it, Garcès showed considerable bravery.

It was not the All Blacks' fault that they were ultimately denied the services of players as dangerous as Ryan Crotty, Waisake Naholo, Rieko Ioane and the brilliant full-back Ben Smith and were forced into a root-and-branch restructuring of their entire back division. They could also play the victimhood card over the late penalty decisions – one by Garcès in the Wellington Test, the other by his countryman Romain Poite in the decider at Eden Park – that handed the Lions a share of the spoils. If, in purely technical terms, Garcès was right in sanctioning the prop Charlie Faumuina for taking out his opposite number Kyle Sinckler in the air, there were sufficient mitigating circumstances to suggest that recent attempts to maximise player safety through law adjustments have fallen prey to a very different kind of law: the one that governs unforeseen consequences. This was followed by the strange business at the last knockings of the deciding Test, when Poite awarded a penalty against the Lions and then changed his mind in favour of a 'deal', resulting in a scrum to the New Zealanders rather than a shot at the sticks. Judging by the expression on the face of Kieran Read, the All Black captain, this was the compromise from hell. Two match-defining decisions in two Tests, both of them against the home side? I love the Lions as much as anyone, and more than most, but I cannot, hand on heart, blame the All Blacks for feeling just a little miffed.

But equally, they had more than enough of all three Tests, and more than enough gilt-edged scoring opportunities, to have rendered isolated penalty decisions null and void. The fact that players as exceptionally gifted as the outside-half Beauden Barrett and the wing Julian Savea, among others, allowed so many points to slip away proved beyond all reasonable doubt that in the very biggest games, pressure tells. And for each and every individual involved in those Test matches, particularly the last, the occasion was as big as it gets. Only a World Cup final can compete and in the All Black mind-set, a

Lions series may in one important way matter more. Why? Because of the 12-year cycle currently in operation. Back in the day, before the isolation of apartheid South Africa and the rise of Australia to stand-alone status in Lions terms, red-shirted visits to New Zealand were not necessarily once-in-a-career events. There were tours to the silver-ferned region in 1959, 1966 and 1971, in 1977 and 1983. Among those great players who had two or more cracks at the most revered of touring sides were Colin Meads, Kel Tremain, Brian Lochore, Tane Norton, Ian Kirkpatrick, Sid Going, Bryan Williams and Andy Haden. The game has changed since then, and with it the schedule. You can name an entire side of truly formidable All Blacks who, because their rugby came to full flower at the wrong moment, never played a Test against the Lions. Think of a pair of wings as good as Jeff Wilson and Jonah Lomu, or half-backs as successful as Andrew Mehrtens and David Kirk, or a back-row combination as potent as Alan Whetton, Josh Kronfeld and Wayne Shelford. For those New Zealanders who happen to be around at the right time and are given the chance, a Lions Test is both something to treasure and something that defines them. Victory is all-important, not least because they almost certainly won't get another go. The 2017 All Blacks will be bitterly disappointed at the outcome of their series, to the point of being distraught.

For the Lions, the opposite must be true. They handled the pressure better than the All Blacks and that, in my book, is an exceptional achievement. The more optimistic among us thought they could be competitive – the squad selected by Gatland showed signs of being very strong physically and temperamentally, and proved to be so – but while they travelled in hope as all touring sides do, nobody gave them a prayer of emerging honours even. Whatever state the All Blacks were in emotionally, the obstacles facing the Lions were far greater and the imponderables with which they were forced to wrestle were incalculably more challenging. No matter how rich the quality of a party, the same questions always kick in. Is there sufficient time to get organised? Will the right combinations emerge naturally, or will they have to be forced together? Which players will prosper in the Lions environment, so different to anything they might previously have experienced, and which ones will go the other way? The fact that there are no immediate answers available means everything is stacked against you. A World Cup final is a pretty big deal, but at least you're starting it alongside a majority of the people you've played alongside, with whom you've shared experiences good and bad, for the previous four years. Here, you're putting your trust in blokes you've only just met and have never shared anything apart from the intense rivalry found on the international field. When Jonny Sexton of Ireland and Owen Farrell of England started the crucial second Test in Wellington at Nos 10 and 12, they were effectively in 'suck it and see' territory, having spent next to no time together. It was, to use the common parlance, a big ask. That's what being a Lion demands of you.

That they were cast in those roles, in those circumstances, was a consequence of Gatland's missteps ahead of the first Test in Auckland.I thought the Lions were really poor in that game. Yes, they fired a couple of shots in the backs. Yes, they scored a truly great 90-metre try – a combination of instinctive counter-attacking work from the Welsh full-back Liam Williams; intelligent continuity and capitalisation work from his countryman Jonathan Davies (one of the players of the tour) and the English wing Daly; and big-energy support from the Irish flanker Sean O'Brien. (A true United Nations effort: I dare say a Scot would also have been involved, had there been one on the field.) What the Lions didn't do, however, was produce the kind of winning rugby they'd employed in two of the biggest pre-Test matches, against the Crusaders in Christchurch and the Maori All Blacks in Rotorua. Instead, they tried to play too much in open field, attempting to take on the All Blacks at their own game and finishing a distant second. To my eyes, they were befuddled, bemused and just a little clueless. It was almost as though they were sent on to the pitch with the wrong message: that Warren had surprised everyone (maybe even himself) by picking a quicksilver back three full of attacking potential but bereft of Lions experience, as opposed to a more functional mix of power and defensive security of the kind he'd employed against the Wallabies in Australia in 2013, and committing himself to a wide game as a consequence. The language he used in the build-up – all that stuff about 'having the courage to play'

– may have come across well on the airwaves and in the public prints, but the upshot was an abandonment of the basics. And unless you do the basics, you're most unlikely to beat New Zealand.

The flipside to all this was that Warren and his coaching team reacted in precisely the right way, tactically and selection-wise, in the wake of defeat. The switch to a Sexton–Farrell axis worked well, as did the changes to the forward pack. I was struck by the parallels with my own Lions tours, to Australia in 1989 and New Zealand in 1993, both under the leadership of Ian McGeechan. In the first of those series, we took a real hiding from the Wallabies in the opening Test in Sydney. For the second, there were significant switches of personnel. I'd like to think my promotion to starting outside-half was the key, but however much I might try, I wouldn't be able to sustain the argument. The new centre pairing of Scott Hastings and a young Jeremy Guscott – what an impact *he* made! – could not have been more effective and the same could be said for the introduction of two super-tough English forwards, the lock Wade Dooley and the flanker Mike Teague, up front. Those new faces were hugely influential in helping us square, and then win, a memorable series.

Something similar happened in '93, not that we were hammered by the All Blacks first up in Christchurch in the way we had been in Sydney. For the second Test in, yes, Wellington – the coincidence is spooky – there was a change at inside centre, Geech taking the dramatic decision to drop

Will Carling, a double Grand Slam-winning England captain, for Scott Gibbs of Wales. He also fell back on his previous trick of replacing Celtic forwards with English ones: in came Brian Moore at hooker, Jason Leonard at tight-head prop and a fresh-faced Martin Johnson in the boilerhouse of the scrum. They made the strongest of contributions to our victory that day, just as Maro Itoje and Sam Warburton did this time around.

Itoje, Warburton, the Sexton–Farrell link . . . these moves in selection (which, as I've already argued, is pretty much the A to Z of coaching, with virtually everything else falling into the 'wallpaper' category) gave the Lions a better balance and the impetus to make a proper fist of it. They rode their luck, of course, and we on this side of the world have to accept that they had an awful lot of luck to ride: even down to 14 men in Wellington, the All Blacks could legitimately ask themselves how they failed to win the series there and then; at 12-6 up at half-time in the decider at Eden Park, they must have wondered how they were not twice as far ahead. But by the same yardstick, we must also give the Lions their due. They showed tenacity, they showed togetherness, they showed will-power, they drilled down so deeply into their reserves of bloody-mindedness that by the end, there was nothing left to mine. When, in the final Test, I saw Farrell kicking his goals under the greatest heat imaginable, even though the rest of his game was a long way out of shape, I could not help thinking back 14 years to Jonny Wilkinson's pressure strikes in the

World Cup final against the Wallabies in Sydney. 'That,' I thought to myself, 'is character. It's not something you can buy. You either have it, or you don't.' Which is why I believe it would have been cruel in the extreme had the Lions lost the series at the death in the way they might have done had Poite not changed his mind on the penalty call. If the referee was wrong in law, the outcome was probably right.

I am quite sure the All Black nation will disagree with that last sentiment, but if they continue to dwell on the injustice of it, they will be dwelling on the wrong thing. The silver-ferned hierarchy would do far better to reflect long and hard on their own frailties and vulnerabilities, the failures in execution that denied them victory in a series that was theirs for the taking. But as a passionate Englishman I would sooner look at things through the prism of red-rose rugby, and the way I see it, what happened in New Zealand in June and July may well have a bearing on what occurs in Japan in 2019, when the World Cup next goes up for grabs. Mako Vunipola, Jamie George, Kyle Sinckler, Maro Itoje, George Kruis, all of them were Test forwards on the Lions tour, and that experience could prove the difference between England competing with the All Blacks and the other leading sides, and beating them. It doesn't end there: had he been fit, Billy Vunipola would have been the Lions' No 8 in the big matches. Owen Farrell, Ben Te'o, Elliot Daly and Anthony Watson also know what it is to play the game of rugby at its limits. The future is certainly bright. Is it also the case that the future is white? There's a chance, for sure, and if it happens, we will surely thank the enduring rugby miracle that is the Lions.

When I'm asked to look back on my own playing days, people often press me to choose between England and the Lions. Which is the greater honour? Which of them means more? My response is always the same: 'England OR the Lions? It's the wrong question, asked for the wrong reasons.' They are two different things and you cannot compare them. In fact, the incomparability is the whole point. They are wholly different productions staged in the same theatre, entirely different forms of music in the same concert hall. Representing the Lions is, if you like, the rugby equivalent of playing golf in the Ryder Cup. How would a Rory McIlroy or a Phil Mickelson choose between winning the Open or the Masters as an individual and winning the Ryder Cup as part of a team – not any old team, either, but a team that will almost certainly not play together a second time? The truth is, you want to do both. As a British rugby player you dream first and foremost of playing for your country. The next step? 'Can I be a Lion having played for my country? Can I wear the jersey and bear the badge?' And then? 'Can I become a Test Lion?' And then? 'Can I win a Test match?' And finally? 'Can I win a Test series?' If you're lucky enough to be fit and healthy and around at the right time, and if you're good enough to be considered for selection, all these things come together. England caps and Lions caps, Grand Slams and World Cups and Lions Test matches, these things are precious. I would not trade my memories of playing in England white for all the riches of the universe. The same goes for my memories of playing in Lions red. Long may the best players in Britain and Ireland be similarly blessed.

ROB ANDREW STATISTICS

ENGLAND

1985

V Romania (Twickenham) W 22-15 – 2 x DG, 4 x Pen (18)
V France (Twickenham) D 9-9 – 1 x DG, 2 x Pen (9)
V Scotland (Twickenham) W 10-7 – 2 x Pen (6)
V Ireland (Dublin) L 10-13 – 2 x Pen (6)
V Wales (Cardiff) L 15-24 – 1 x Con, 1 x DG, 2 x Pen (11)

P 5 W 2 D 1 L 2
1 x Con, 4 x DG, 12 x Pen (50)

1986

V Wales (Twickenham) W 21-18 – 1 x DG, 6 x Pen (21)
V Scotland (Edinburgh) L 6-33 – 2 x Pen (6)
V Ireland (Twickenham) W 25-20 – 3 x Con, I x Pen (9)
V France (Paris) L 10-29

P 4 W 2 D 0 L 2
3 x Con, 1 x DG, 9 x Pen (36)

1987

V Ireland (Dublin) L 0-17
V France (Twickenham) L 15-19 – 1 x DG (3)
V Wales (Cardiff) – L 12-19
V Japan (Sydney, WC, rep) W 60-7
V United States (Sydney, WC) W 34-6

P 5 (1 x rep) W 2 D 0 L 3
I x DG (3)

1988

V Scotland (Edinburgh) W 9-6 – 1 x DG (3)
V Ireland (Twickenham) W 35-3 – 3 x Con (6)
V Ireland (Dublin, Mill) W 21-10
V Australia (Brisbane) L 16-22
V Australia (Sydney) L 8-28
V Fiji (Suva, FB) W 25-12
V Australia (Twickenham) W 28-19

P 7 W 5 D 0 L 2
3 x Con, 1 x DG (9)

1989

V Scotland (Twickenham) D 12-12 – 2 x Pen (6)
V Ireland (Dublin) W 16-3 – 1 x Con, 2 x Pen (8)
V France (Twickenham) W 11-0 – 1 x Pen (3)
V Wales (Cardiff) L 9-12 – 1 x DG, 2 x Pen (9)
V Romania (Bucharest) W 58-3 – 1 DG (3)
V Fiji (Twickenham) W 58-23 – 1 x Con (2)

P 6 W 4 D 1 L 1
2 x Con, 2 x DG, 7 x Pen (31)

1990

V Ireland (Twickenham) W 23-0
V France (Paris) W 26-7
V Wales (Twickenham) W 34-6
V Scotland (Edinburgh) L 7-13
V Argentina (Twickenham) W 51-0

P 5 W 4 D 0 L 1

1991

V Wales (Cardiff) W 25-6
V Scotland (Twickenham) W 21-12
V Ireland (Dublin) W 16-7
V France (Twickenham) W 21-19 – 1 x DG (3)
V Fiji (Suva) W 28-12 – 1 x Try, 2 x DG (6)
V Australia (Sydney) L 15-40
V New Zealand (Twickenham, WC) L 12-18 – 1 x DG (3)
V United States (Twickenham, WC) W 37-9
V Italy (Twickenham, WC) W 36-6
V France (Paris, WC) W 19-10
V Scotland (Edinburgh, WC) W 9-6 – 1 x DG (3)
V Australia (Twickenham, WC) L 6-12

P 12 W 9 D 0 L 3
1 x Try, 5 x DG (15)

1992

V Scotland (Edinburgh) W 25-7
V Ireland (Twickenham) W 38-9
V France (Paris) W 31-13
V Wales (Twickenham) W 24-0
V Canada (Wembley) W 26-13
V South Africa (Twickenham) W 33-16

P 6 W 6 D 0 L 0

1993

V France (Twickenham) W 16-15
V Wales (Cardiff) L 9-10
V New Zealand (Twickenham) W 15-9 – 1 x DG (3)

P 3 W 2 D 0 L 1

1 x DG (3)

1994

V Scotland (Edinburgh) W 15-14
V Ireland (Twickenham) L 12-13
V France (Paris) W 18-14 – 1 x DG, 5 x Pen (18)
V Wales (Twickenham) W 15-8 – 1 x Con, 1 x Pen (5)
V South Africa (Pretoria) W 32-15 – 1 x Try, 2 x Con, 1 x DG, 5 x Pen (27)
V South Africa (Cape Town) L 9-27 – 3 x Pen (9)
V Romania (Twickenham) W 54-3 – 6 x Con, 4 x Pen (24)
V Canada (Twickenham) W 60-19 – 6 x Con, 6 x Pen (30)

P 8 W 6 D 0 L 2

1 x Try, 15 x Con, 2 x DG, 24 x Pen (113)

1995

V Ireland (Dublin) W 20-8 – 1 x Con, 1 x Pen (5)
V France (Twickenham) W 31-10 – 2 x Con, 4 x Pen (16)
V Wales (Cardiff) W 23-9 – 1 x Con, 2 x Pen (8)
V Scotland (Twickenham) W 24-12 – 1 x DG, 7 x Pen (24)
V Argentina (Durban, WC) W 24-18 – 2 x DG, 6 x Pen (24)
V Italy (Durban, WC) W 27-20 – 1 x Con, 5 x Pen (17)
V Australia (Cape Town, WC) W 25-22 – 1 x Con, 1 x DG, 5 x Pen (20)
V New Zealand (cape Town, WC) L 29-45 – 3 x Con, 1 x Pen (9)
V France (Pretoria, WC) L 9-19 – 3 x Pen (9)

P 9 W 7 D 0 L 2

9 x Con, 4 x DG, 34 x Pen (132)

1997

V Wales (Cardiff, rep) W 34-13

P 1 (1 x rep) W 1 D 0 L 0

Played 71 (two x rep, one x substituted, one x full-back)
Won 50
Drawn 2
Lost 19
Tries x 2 (9)
Conversions x 33 (66)
Drop goals x 21 (63)
Penalties x 86 (258)
Total points: 396

BRITISH AND IRISH LIONS

1989

V Queensland (Brisbane, rep) W 19-15
V Queensland B (Cairns) W 30-6 – 4 x Con
V ACT (Canberra) W 41-25
V Australia (Brisbane, second Test) W 19-12 – 1 x Con, 1 x DG, 1 x Pen (8)
V Australia (Sydney, third Test) W 19-18
V ANZAC XV (Brisbane, rep) W 19-15

V France (Paris, non-official) W 29-27 – I x Try, 1 x DG (7)

1993

V North Harbour (Auckland) W 29-13 – 1 x Try (5)
V Canterbury (Christchurch) W 28-10 – 1 x Try, 1 x Con, 1 x DG, 1 x Pen (13)
V Southland (Invercargill) W 34-16
V New Zealand (Christchurch, first Test) L 18-20
V Auckland (Auckland) L 18-23 – 1 x Pen (3)
V New Zealand (Wellington, second Test) W 20-7 – 1 x DG (3)
V New Zealand (Auckland, third Test) L 13-30

ALL MATCHES

Played 14 (two x rep)
Won 11
Drawn 0
Lost 3
Tries x 3 (14)
Conversions x 6 (12)
Drop goals x 4 (12)
Penalties x 3 (9)
Total points: 47

Tests

Played 6
Won 4
Drawn 0
Lost 2
Tries x 1 (4)
Conversions x 1 (2)
Drop goals x 3 (9)
Penalties x 1 (3)

CLUB RECORD IN THE PROFESSIONAL ERA

1996/97

Newcastle Falcons (runners-up)
National Division 2
Played 22
Won 19
Drawn 1
Lost 2
Tries x 7
Conversions x 95
Drop goals x 1
Penalties x 23
Total points: 297

1997/98

Newcastle Falcons
Premiership (champions)
Played 22
Won 19
Drawn 0
Lost 3
Tries x 6
Conversions x 44
Drop goals x 1
Penalties x 35
Total points 226

European Challenge Cup (semi-finalists)
Played 8
Won 6

Lost 2
Tries x 1
Conversions x 5
Drop goals x 0
Penalties x 6
Total points 33

1998/99

Newcastle Falcons
Premiership (8th)
Played 21
Won 11
Drawn 0
Lost 10
Tries x 1
Conversions x 6
Drop goals x 0
Penalties x 3
Total points 26

1999/2000

Newcastle Falcons
Premiership (9th)
Played 1
Won 0
Drawn 0
Lost 1
Tries x 0
Conversions x 0
Drop goals x 0
Penalties x 0
Total points 0

PHOTO ACKNOWLEDGEMENTS

The author and publisher would like to thank the following for permission to reproduce photographs:

Section 1
David Rogers/Getty Images, COLORSPORT/Stuart MacFarlane, , Dave Rogers / Allsport/Getty Images, COLORSPORT/Colin Elsey, Offside Sports Photography, COLORSPORT/Colin Elsey, COLORSPORT/Andrew Cowie, COLORSPORT/ Stewart Fraser, David Rogers/Getty Images, COLORSPORT/Steve Bardens, COLORSPORT/Steve Bardens, COLORSPORT/Stuart MacFarlane, COLORSPORT/Richard Hume, Jamie McDonald/Getty Images, COLORSPORT/Andrew Cowie

Section 2
David Rogers/Getty Images, COLORSPORT/Andrew Cowie, Dave Rogers / Allsport/Getty Images, COLORSPORT/Kieran Galvin, COLORSPORT/Matthew Impey, Matthew Lewis/Getty Images, David Rogers/Getty Images, Laurence Griffiths/Getty Images, COLORSPORT/Matthew Impey, COLORSPORT/ Andrew Cowie, David Rogers/Getty Images, David Davies/PA Archive/PA Images, David Rogers/Getty Images, COLORSPORT/Kieran Galvin

Section 3
Warren Little/Getty Images, COLORSPORT/Paul Zammit Cutajar, COLORSPORT/Andrew Cowie, David Rogers/Getty Images, COLORSPORT/ Andrew Cowie, Gareth Fuller/PA Wire/PA Images, COLORSPORT/Andrew Cowie, Livesey/Getty Images, COLORSPORT/Dan Rowley, COLORSPORT/ Andrew Cowie

Section 4
COLORSPORT/Andrew Cowie, COLORSPORT/David Gibson, COLORSPORT/ Winston Bynorth, RFU/The RFU Collection via Getty Images, David Rogers/ Getty Images, COLORSPORT/Andrew Cowie, bottom: COLORSPORT/Ken Sutton, COLORSPORT/Lynne Cameron, bottom Photosport / Offside Sports Photography, COLORSPORT/Lynne Cameron, COLORSPORT/Andrew Cowie

INDEX

INDEX

INDEX